THE NATIONAL ACADEMIES
KECK FUTURES INITIATIVE

SEEING THE FUTURE WITH IMAGING SCIENCE

INTERDISCIPLINARY RESEARCH TEAM SUMMARIES

Conference
Arnold and Mabel Beckman Center
Irvine, California
November 16-19, 2010

THE NATIONAL ACADEMIES PRESS
Washington, D.C.
www.nap.edu

THE NATIONAL ACADEMIES PRESS   500 Fifth Street, N.W.   Washington, DC 20001

NOTICE: The Interdisciplinary Research (IDR) team summaries in this publication are based on IDR team discussions during the National Academies Keck *Futures Initiative* Conference on Imaging Science held at the Arnold and Mabel Beckman Center in Irvine, California, November 16-19, 2010. The discussions in these groups were summarized by the authors and reviewed by the members of each IDR team. Any opinions, findings, conclusions, or recommendations expressed in this publication are those of the IDR teams and do not necessarily reflect the view of the organizations or agencies that provided support for this project. For more information on the National Academies Keck *Futures Initiative* visit www.keckfutures.org.

Funding for the activity that led to this publication was provided by the W. M. Keck Foundation. Based in Los Angeles, the W. M. Keck Foundation was established in 1954 by the late W. M. Keck, founder of the Superior Oil Company. In recent years, the Foundation has focused five broad areas: Science and Engineering research; Undergraduate Education; Medical Research; and Southern California. Each grant program invests in people and programs that are making a difference in the quality of life, now and for the future. For more information visit www.wmkeck.org.

International Standard Book Number-13: 978-0-309-20906-9
International Standard Book Number-10: 0-309-20906-4

Additional copies of this report are available from the National Academies Press, 500 Fifth Street, N.W., Lockbox 285, Washington, DC 20055; (800) 624-6242 or (202) 334-3313 (in the Washington metropolitan area); Internet, http://www.nap.edu.

Copyright 2011 by the National Academy of Sciences. All rights reserved.

Printed in the United States of America

THE NATIONAL ACADEMIES
Advisers to the Nation on Science, Engineering, and Medicine

The **National Academy of Sciences** is a private, nonprofit, self-perpetuating society of distinguished scholars engaged in scientific and engineering research, dedicated to the furtherance of science and technology and to their use for the general welfare. Upon the authority of the charter granted to it by the Congress in 1863, the Academy has a mandate that requires it to advise the federal government on scientific and technical matters. Dr. Ralph J. Cicerone is president of the National Academy of Sciences.

The **National Academy of Engineering** was established in 1964, under the charter of the National Academy of Sciences, as a parallel organization of outstanding engineers. It is autonomous in its administration and in the selection of its members, sharing with the National Academy of Sciences the responsibility for advising the federal government. The National Academy of Engineering also sponsors engineering programs aimed at meeting national needs, encourages education and research, and recognizes the superior achievements of engineers. Dr. Charles M. Vest is president of the National Academy of Engineering.

The **Institute of Medicine** was established in 1970 by the National Academy of Sciences to secure the services of eminent members of appropriate professions in the examination of policy matters pertaining to the health of the public. The Institute acts under the responsibility given to the National Academy of Sciences by its congressional charter to be an adviser to the federal government and, upon its own initiative, to identify issues of medical care, research, and education. Dr. Harvey V. Fineberg is president of the Institute of Medicine.

The **National Research Council** was organized by the National Academy of Sciences in 1916 to associate the broad community of science and technology with the Academy's purposes of furthering knowledge and advising the federal government. Functioning in accordance with general policies determined by the Academy, the Council has become the principal operating agency of both the National Academy of Sciences and the National Academy of Engineering in providing services to the government, the public, and the scientific and engineering communities. The Council is administered jointly by both Academies and the Institute of Medicine. Dr. Ralph J. Cicerone and Dr. Charles Vest are chair and vice chair, respectively, of the National Research Council.

www.national-academies.org

THE NATIONAL ACADEMIES KECK *FUTURES INITIATIVE* IMAGING SCIENCE STEERING COMMITTEE

FAROUK EL-BAZ (Chair) (NAE), Research Professor and Director, Center for Remote Sensing, Boston University
NANCY C. ANDREASEN (IOM), Andrew H. Woods Chair of Psychiatry, University of Iowa Hospitals and Clinics
HARRISON BARRETT, Regents Professor of Radiology, Regents Professor of Optical Sciences, Regents Professor of Applied Mathematics, University of Arizona
FLOYD E. BLOOM (NAS/IOM), Professor Emeritus, Department of Molecular and Integrative Neuroscience, The Scripps Research Institute
RITA R. COLWELL (NAS), Distinguished University Professor, Center for Bioinformatics & Computational Biology, University of Maryland
CHARLES ELACHI (NAE), Director, Jet Propulsion Laboratory
XIAOPING HU, Professor and Georgia Research Alliance Eminent Scholar in Imaging; Director, Biomedical Imaging Technology Center; Scientific Director, Center for Systems Imaging, Coulter Department of Biomedical Engineering; Georgia Tech; Emory University
JOANNE O. ISHAM, Aurora Group
VICTORIA MORGAN, Assistant Professor of Radiology and Radiological Sciences, Assistant Professor of Biomedical Engineering, Vanderbilt University
BRIAN A. WANDELL (NAS), Isaac and Madeline Stein Family Professor, Department of Psychology, Stanford University

Staff

KENNETH R. FULTON, Executive Director
KIMBERLY A. SUDA-BLAKE, Senior Program Director
ANNE HEBERGER MARINO, Senior Evaluation Associate
CRISTEN KELLY, Program Associate
RACHEL LESINSKI, Program Associate

The National Academies Keck *Futures Initiative*

THE NATIONAL ACADEMIES KECK *FUTURES INITIATIVE*

The National Academies Keck *Futures Initiative* was launched in 2003 to stimulate new modes of scientific inquiry and break down the conceptual and institutional barriers to interdisciplinary research. The National Academies and the W. M. Keck Foundation believe that considerable scientific progress will be achieved by providing a counterbalance to the tendency to isolate research within academic fields. The *Futures Initiative* is designed to enable scientists from different disciplines to focus on new questions, upon which they can base entirely new research, and to encourage and reward outstanding communication between scientists as well as between the scientific enterprise and the public.

The *Futures Initiative* includes three main components:

Futures Conferences

The *Futures* Conferences bring together some of the nation's best and brightest researchers from academic, industrial, and government laboratories to explore and discover interdisciplinary connections in important areas of cutting-edge research. Each year, some 150 outstanding researchers are invited to discuss ideas related to a single cross-disciplinary theme. Participants gain not only a wider perspective but also, in many instances, new insights and techniques that might be applied in their own work. Additional pre- or post-conference meetings build on each theme to foster further communication of ideas.

Selection of each year's theme is based on assessments of where the intersection of science, engineering, and medical research has the greatest potential to spark discovery. The first conference explored *Signals, Decisions, and Meaning in Biology, Chemistry, Physics, and Engineering.* The 2004 conference focused on *Designing Nanostructures at the Interface between Biomedical and Physical Systems.* The theme of the 2005 conference was *The Genomic Revolution: Implications for Treatment and Control of Infectious Disease.* In 2006 the conference focused on *Smart Prosthetics: Exploring Assistive Devices for the Body and Mind.* In 2007 the conference explored *The Future of Human Healthspan: Demography, Evolution, Medicine and Bioengineering.* In 2008 the conference focused on *Complex Systems.* The 2009 conference explored *Synthetic Biology: Building on Nature's Inspiration.* The 2010 conference focused on *Seeing the Future with Imaging Science* and the 2011 conference will focus on Ecosystem Services.

Futures Grants

The *Futures* Grants provide seed funding to *Futures* Conference participants, on a competitive basis, to enable them to pursue important new ideas and connections stimulated by the conferences. These grants fill a critical missing link between bold new ideas and major federal funding programs, which do not currently offer seed grants in new areas that are considered risky or exotic. These grants enable researchers to start developing a line of inquiry by supporting the recruitment of students and postdoctoral fellows, the purchase of equipment, and the acquisition of preliminary data—which in turn can position the researchers to compete for larger awards from other public and private sources.

NAKFI Communications

The Communication Awards are designed to recognize, promote, and encourage effective communication of science, engineering, medicine, and/or interdisciplinary work within and beyond the scientific community. Each year the *Futures Initiative* awards $20,000 prizes to those who have advanced the public's understanding and appreciation of science, engineering, and/or medicine. The awards are given in four categories: books, magazine/newspaper, online, and film/radio/TV. The winners are honored during a ceremony in the fall in Washington, DC.

NAKFI cultivates science writers of the future by inviting graduate students from science writing programs across the country to attend the conference and develop IDR team discussion summaries and a conference overview for publication in this book. Students are selected by the department director or designee, and they prepare for the conference by reviewing the webcast tutorials and suggested reading and selecting an IDR team in which they would like to participate. Students then work with NAKFI's science writing scholar mentor to finalize their reports following the conferences.

Facilitating Interdisciplinary Research Study

During the first 18 months of the Keck *Futures Initiative*, the Academies undertook a study on facilitating interdisciplinary research. The study examined the current scope of interdisciplinary efforts and provided recommendations as to how such research can be facilitated by funding organizations and academic institutions. *Facilitating Interdisciplinary Research* (2005) is available from the National Academies Press (www.nap.edu) in print and free PDF versions.

About the National Academies

The National Academies comprise the National Academy of Sciences, the National Academy of Engineering, the Institute of Medicine, and the National Research Council, which perform an unparalleled public service by bringing together experts in all areas of science and technology, who serve as volunteers to address critical national issues and offer unbiased advice to the federal government and the public. For more information, visit www.nationalacademies.org.

About the W. M. Keck Foundation

Based in Los Angeles, the W. M. Keck Foundation was established in 1954 by the late W. M. Keck, founder of the Superior Oil Company. The Foundation's grant making is focused primarily on pioneering efforts in the areas of Science and Engineering; Undergraduate Education; Medical Research; and Southern California. Each grant program invests in people and programs that are making a difference in the quality of life, now and in the future. For more information visit www.wmkeck.org.

National Academies Keck *Futures Initiative*
100 Academy, 2nd Floor
Irvine, CA 92617
949-721-2270 (Phone)
949-721-2216 (Fax)
www.keckfutures.org

Preface

At the National Academies Keck *Futures Initiative* Conference on Imaging Science, participants were divided into 14 interdisciplinary research teams. The teams spent nine hours over two days exploring diverse challenges at the interface of science, engineering, and medicine. The composition of the teams was intentionally diverse, to encourage the generation of new approaches by combining a range of different types of contributions. The teams included researchers from science, engineering, and medicine, as well as representatives from private and public funding agencies, universities, businesses, journals, and the science media. Researchers represented a wide range of experience—from postdoc to those well established in their careers—from a variety of disciplines that included science and engineering, medicine, physics, biology, math/computer science, and behavioral science.

The teams needed to address the challenge of communicating and working together from a diversity of expertise and perspectives as they attempted to solve a complicated, interdisciplinary problem in a relatively short time. Each team decided on its own structure and approach to tackle the problem. Some teams decided to refine or redefine their problems based on their experience.

Each team presented two brief reports to all participants: (1) an interim report on Thursday to debrief on how things were going, along with any special requests; and (2) a final briefing on Friday, when each team:

- Provided a concise statement of the problem;
- Outlined a structure for its solution;
- Identified the most important gaps in science and technology and recommended research areas needed to attack the problem; and
- Indicated the benefits to society if the problem could be solved.

Each IDR team included a graduate student in a university science writing program. Based on the team interaction and the final briefings, the students wrote the following summaries, which were reviewed by the team members. These summaries describe the problem and outline the approach taken, including what research needs to be done to understand the fundamental science behind the challenge, the proposed plan for engineering the application, the reasoning that went into it, and the benefits to society of the problem solution. Because of the popularity of some topics, two or three teams were assigned to explore the subjects.

Seven webcast tutorials were launched throughout the summer to help bridge the gaps in terminology used by the various disciplines. Participants were encouraged to view all of the tutorials prior to the November conference.

Contents

Conference Summary   1

IDR TEAM SUMMARIES

Team 1: Develop a method to integrate neuroimaging technologies at different length and time scales.   5
    IDR Team Summary, Group A   7
    IDR Team Summary, Group B   12

Team 2: Identify the mathematical and computational tools that are needed to bring recent insights from theoretical image science and rigorous methods of task-based assessment of image quality into routine use in all areas of imaging.   21
    IDR Team Summary, Group A   24
    IDR Team Summary, Group B   27

Team 3: Develop and validate new methods for detecting and classifying meaningful changes between two images taken at different times or within temporal sequences of images.   35
    IDR Team Summary, Group A   37
    IDR Team Summary, Group B   42
    IDR Team Summary, Group C   46

Team 4: Develop a telescope or starshade that would allow
planetary systems around neighboring stars to be imaged.   53

Team 5: How can we extend the domain of adaptive optics and
adaptive imaging to new application, and how can we objectively
compare adaptive and non-adaptive approaches to specific
imaging problems?   61

Team 6: What are the tools and validation methods required to
develop clinically useful non-invasive imaging biomarkers of
psychiatric disease?   71

Team 7: Find novel ways to use imaging methods to improve
the treatment of diseases.   79
    IDR Team Summary, Group A   83
    IDR Team Summary, Group B   88
    IDR Team Summary, Group C   94

Team 8: Develop image-specialized database tools for data
stewardship and system design in large-scale applications.   101

APPENDIXES

List of Webcast Tutorials   109
Agenda   111
Participants   115

To view the webcast tutorials or conference presentations,
please visit our website at www.keckfutures.org.

Conference Summary

Tia Ghose

Imaging science has the power to illuminate regions as remote as distant galaxies, and as close to home as our own bodies. Everything from medicine to carbon sequestration is the potential beneficiary of masses of new data, and researchers are struggling to make sense of it all and communicate its meaning to other researchers. Many of the disciplines that can benefit from imaging share common technical problems. Yet researchers often develop ad hoc methods for solving individual tasks without building broader frameworks that could address many scientific problems.

At the 2010 National Academies Keck *Futures Initiative* Conference on Imaging Science, researchers were asked to find a common language and structure for developing new technologies, processing and recovering images, mining imaging data, and visualizing it effectively. A common theme emerged: how do you find what matters in a sea of information that is varied, incomplete, or simply monstrously large in size and scope? This problem is particularly tricky because scientists may know the underlying truth that they are seeking, but are often unsure how it will look in a certain imaging technique. For some, the task was picking out the dim light of a tiny planet obscured by a sun billions of times brighter. Others aimed to mine satellite images to track tiny specks of land that are clear cut in a Brazilian rainforest. Still others hoped to turn the power of imaging inward, to find hidden tumors or signs of Alzheimer's disease decades before people show symptoms.

The *Keck Futures Initiative* highlighted Imaging Science to spur researchers working on similar problems across disciplines to create common solutions and language. It brought researchers from academia, industry, and government together into 14 Interdisciplinary Research (IDR) teams to develop creative thought outside the confines of any individual area of expertise.

IDR teams 1A&B grappled with how to integrate images of the brain with tools like MRI, PET scans, EEG, or microscopy, which each operate on different time and length scales. Some can capture signaling molecules that are just a few hundred nanometers, others map neurons that are tens of micrometers, while still others track the electrical impulses coursing through our brains. But there is no framework for combining this grab-bag of techniques to say how signaling molecules relate to gray matter, or how an MRI scan showing shrinkage in an Alzheimer's disease–riddled brain corresponds to the lower oxygen usage shown on a PET scan. Some members quickly realized that to integrate data from the tiny to the large, you need to perform imaging using many devices at once. They proposed doing a panel of imaging tests on animals and humans, developing models of how those images related to brain function and to each other.

Teams answering challenge 2 discussed whether it was possible to create overall metrics to evaluate an imaging system's performance. One team determined that no metric will be useful unless it can account for, and adjust to, the person interpreting an image. They developed the idea of creating a system that was tailored to an individual reader's biases and preferences. They also emphasized that tasks like picking out the tiny tumor in an X-ray rely on key contextual information that isn't available in the images themselves, and that good metrics need to account for this information. For instance, radiologists use context like the patient's history and symptoms to hone in on the areas to scan.

Researchers in team 3 aimed to detect meaningful changes between two images. Some tasks, like mapping deforestation, rely on grainy satellite images that are often altered by cloud cover, rainy days, or snow. Although there are many powerful algorithmic tools available, most researchers develop ad hoc solutions for these tasks and don't really share their approaches with others. One group decided that a web-based tutorial inspired by the much-loved *Numerical Recipes* textbook could be combined with a grand challenge competition to help standardize the toolsets researchers use in image processing. Another group decided that tracking a sequence of images over time, rather than just two images, would allow them to identify more meaningful trends in the data.

IDR group 4 was charged with finding exoplanets that circled distant suns. The physical devices needed to find these planets are already being developed, so the group focused on building image processing algorithms. This task is difficult because most of the exoplanets found so far haven't looked anything like the predictions, so astronomers aren't quite sure what they should even be looking for. They noted that an algorithm should account for the disturbances in the image caused by filtering out the starlight, should distinguish the blue dot of an exoplanet from streaks of star light, and should pick out the planet's motion as it orbited its sun. They also hoped to adapt their observational methods, so that, instead of spending a fixed amount of time monitoring each portion of the sky, they could spend more time gathering light from promising areas while quickly moving on from less promising ones.

Although adaptive optics has already revolutionized astronomy, team 5 aimed to extend the approach to other arenas. In the classic adaptive optics set-up, light is sent out through a medium, and the altered wave is recorded; a lens can then correct for that aberration by altering its shape with deformable mirrors. The researchers decided that adaptive optics could be especially useful for peering inside the body. They envisioned expanding the technology from two dimensions to create volumetric imaging—looking at hearts, lungs, and brains in 3-D. They also thought the technique could be expanded to peer through tissue that usually scatters light waves, so that fuzzy objects inside the cell could be seen more clearly.

Team 6 focused on finding the robust markers of psychiatric diseases like autism spectrum disorder and schizophrenia. Although these diseases are usually diagnosed by their symptoms rather than a definitive test, the underlying structure and function of the brain is at the root of these conditions. Thus, imaging techniques like PET and MRI should be able to reveal the brain's dysfunction. Unfortunately, all of these techniques can mistake healthy brains for diseased ones, so the team decided a panel of multiple markers would be needed to accurately find signs of disease. They also emphasized that, because behavior is the hallmark of these diseases, new techniques to monitor people in more natural environments, such as gaze tracking and portable electrical activity readers, could be developed to strengthen some of these biomarkers.

Team 7's challenge was to incorporate several imaging methods to streamline disease treatment and diagnosis. The team quickly focused on cancer and imagined a future in which MRI, PET, CT, and other diagnostic imaging could be integrated into one, multipurpose device to facilitate disease diagnosis and targeted treatment. One group imagined 3-D goggles that could continuously scan people's retinas for signs of metastasis in their blood cells, instead of requiring patients to come in every few months for an invasive blood test. As one team member noted, "Who wouldn't want to watch a 3-D movie with their family and decide if you have disease at the same time?"

Team 8 aimed to develop better architecture to store, curate, and make sense of the data deluge from imaging science. Currently, images collected in biological disciplines, including neuroscience, are stored in different formats, come from a constantly changing array of instruments, and look at different underlying physical phenomena. In addition, databases work well when you know what you are looking for, but they currently lack the tools to explore the data in images in a less directed manner. The team envisioned developing standards for data searches and also imagined an architecture that supports image processing and operates as part of the database. The team developed a concept of exploratory tools that let people collect and analyze image data and imagined using machine learning to anticipate what someone is seeking, even when they're not quite sure themselves.

At the close of the conference, many researchers noted how valuable it was to speak with people outside their disciplines. Although the current field of imaging science is full of many different languages, for just a few days, researchers spoke a common language. With the avalanche of imaging data expected in the coming years, an ability to tackle broader problems systematically and to find meaning in the madness will only become more important.

IDR Team Summary 1

Develop a method to integrate neuroimaging technologies at different length and time scales.

CHALLENGE SUMMARY

The neurosciences and medical imaging have produced a diverse array of technologies that measure neural structures and signals. These methods acquire information over a wide range of length and temporal scales, ranging from magnetic resonance (MR) and electroencephalogram (EEG) data in the intact human brain (at the scale of centimeters) to electron microscopy and two-photon imaging at the sub-micron scale. Each of these imaging technologies contributes different but ultimately related understanding of the brain's neural circuitry. There is fertile ground for the application of integration techniques; however, currently there is risk of dividing the data acquired using these different modalities into segregated fields. The challenge is to integrate the measurements obtained using these different technologies at different length and time scales. This must be possible because, in the end, all of these measurements provide information about the same basic neural circuitry. Combining the data across the variety of imaging technologies requires individuals and tools that are capable of understanding the neural circuitry and signaling; we need to develop a model that can integrate the data and the implications of these different measurements into a coherent whole.

Following are several examples of how progress might be made. First, it would be important to understand and quantify the relationship among key elements of neural signaling—such as resetting ion channel potentials, transmitter recycling, action potentials, sub-threshold synaptic potentials, glial signaling—and global signals such as fMRI (functional magnetic resonance imaging), EEG (electroencephalogram) and MEG (magnetoencephalography). Second, it would be important to understand the implications of the dendritic and axonal arbors for the mean electrical field and its several frequency components (gamma band, alpha band, and so forth) as measured in clinical and scientific studies in EEG and MEG. Third, it would be important to understand the relationship between neurotransmitter concentrations, such as aminolbutric acid (GABA) density measured using MR spectroscopy, and circuit properties, such as the peak oscillation and coherence bands. Finally, it would be important to have the ability to generate a computational model of a circuit with specific anatomy so that the simultaneous prediction of the fMRI signal, the EEG signal, and the two-photon calcium images from this same circuit is possible given a particular input.

To systematically understand the relationship of data at different scales, it is necessary to establish theories and mathematical models to link the data and to validate these models with experimental data from in vitro settings and in vivo settings with animal models and human subjects. For applications to disease, it is also necessary to include pathological alterations of these models. Although there have been ad hoc efforts to combine data from different modalities, a systematic approach—which may lead to groundbreaking methodologies and science—is lacking.

Key Questions

- How do we establish a common computational language that might be used by investigators using these diverse technologies to measure neural circuitry and neural signals?
- Can we identify some key model systems that would serve as a fruitful environment for combining these techniques? Can these be human, or does the basic work have to be done in animal systems?
- How do we educate investigators who are principally involved in one technology—say fMRI or two-photon calcium imaging—in the biophysics and modeling techniques that would allow them to understand the related fields and contribute to the complete modeling effort?

Reading

Appelbaum LG, Wade AR, Vildavski VY, Pettet MW, Norcia AM. Cue-invariant networks for figure and background processing in human visual cortex. *J Neurosci* 2006 Nov 8;26(45):11695-708. PMID: 17093091. Accessed online June 15, 2010.

Lichtman JW, Livet J, Sanes JR. A technicolour approach to the connectome. *Nat Rev Neurosci* 2008;9:417-422. Accessed online June 15, 2010.

Logothetis NK, Wandell BA. Interpreting the BOLD signal. *Annu Rev Physiol* 2004;66:735-69. Accessed online June 15, 2010.

Logothetis NK. What we can do and what we cannot do with fMRI. *Nature* 2008 Jun 12;453(7197):869-78. Review. PMID: 18548064. Accessed online June 15, 2010.

Ohki K, Chung S, Ch'ng YH, Kara P, Reid RC. Functional imaging with cellular resolution reveals precise micro-architecture in visual cortex. *Nature* 2005 Feb 10;433(7026):597-603. Epub 2005 Jan 19. Accessed online June 15, 2010.

Sharon D, Hämäläinen MS, Tootell RB, Halgren E, Belliveau JW. The advantage of combining MEG and EEG: comparison to fMRI in focally stimulated visual cortex. *Neuroimage* 2007 Jul 15;36(4):1225-35. Epub 2007 Apr 19. PMID: 17532230. Accessed online June 15, 2010.

Sporns O, Tononi G, Kötter R. The human connectome: a structural description of the human brain. *PLoS Comput Biol* 2005 Sep;1(4):e42. Accessed online June 15, 2010.

Because of the popularity of this topic, two groups explored this subject. Please be sure to review the second write-up, which immediately follows this one.

IDR TEAM MEMBERS—GROUP A

- Richard A. Baird, National Institutes of Health
- Randy A. Bartels, Colorado State University
- DuBois Bowman, Emory University
- Joseph E. Burns, University of California, Irvine
- J. Lawrence Marsh, University of California, Irvine
- Gregory M. Palmer, Duke University
- Steven G. Potkin, University of California, Irvine
- Suzanne Scarlata, Stony Brook University
- Mercedes Talley, W. M. Keck Foundation
- Paul Vaska, Brookhaven National Laboratory
- Lihong V. Wang, Washington University in St. Louis
- Jordan Calmes, Massachusetts Institute of Technology

IDR TEAM SUMMARY—GROUP A

*Jordan Calmes, NAKFI Science Writing Scholar,
Massachusetts Institute of Technology*

The current state of neuroimaging is reminiscent of the classic story of six blind men describing an elephant. One of the men has access to the elephant's tusk, and concludes that an elephant is like a spear. The man standing right next to him, touching the trunk instead, decides the elephant must be like a snake. Each of the blind men has a detailed but limited view of their subject, and although each of them has access to factual information, none of them can claim a complete knowledge of the elephant.

The blind men are lucky, in that they only want to describe the outside of the elephant, whereas neuroscientists have to work from the systems level all the way down to the cellular level. A researcher looking at a magnetic resonance image (MRI) of a complete brain (at a centimeter scale) and a researcher looking at single-cell connections within that brain (at a submicron scale) have a gigantic barrier to overcome if they hope to collaborate.

As neuroimaging techniques have improved, there has been movement toward integrating various techniques so that one will reveal a more complete picture of the brain. Although this seems like a huge task, it is not impossible. Each technique runs at a different spatial and time scale, but they all measure the same basic circuitry.

Combining data from different technologies will require researchers and tools capable of understanding that basic neural circuitry in great depth so that they can create a model that can integrate the various measurements into data that makes sense to the investigator. First, the researchers will need an in-depth understanding of the relationships between key elements of neural signaling, processes like action potentials and glial signals, and techniques like functional magnetic resonance imaging (fMRI) or electroencephalography (EEG). Second, the team will need to understand the effects of signals from different types of nerve cells on the brain's electric field. Third, the team will need to understand the relationship between the properties of neural circuits and the concentrations of different chemicals in the brain. Finally, someone must be able to generate a computational model of a circuit that can predict fMRI signals and EEG signals at the same time.

To achieve these tasks, researchers would first need to establish a common computational language. Then, they would have to identify human or animal model systems that could be used for experimentation. Finally, someone would have to develop a program to educate experts who work with one technology on the other applications. These were the major questions that Interdisciplinary Research team 1A explored during the conference.

Defining the Challenge

Team 1A first worked to outline the advantages to integrating neuroimaging techniques. They all agreed that integrating the technologies would lead to a whole that was greater than the sum of the parts. Integrating imaging technologies across spatial and temporal scales should result in something that the simple ping-pong from one modality to another could not achieve. Already, there is quite a bit of interaction between people studying the brain at different scales. People use microscale techniques to develop new macroscale techniques in animals, which are then used to develop new techniques for use in humans, which lead to new questions that feed back into microscale research on animals. None of that is new.

The team had more trouble deciding how to integrate the technologies. Should the data be collected at the same time? Could data be better integrated with diagnosis?

More importantly, why should anyone go to the trouble? What is it that we could learn about the brain by integrating neuroimaging techniques?

Most methods of brain imaging use indirect contrasts. Often, scientists are not sure exactly what their tools are measuring within the brain. Linking modalities with indirect contrasts to those with direct contrasts, fMRI with EEG for example, could help improve our understanding of what the indirect contrasts are measuring.

Microscopic imaging can enable optimization of macroscopic imaging. Macroscopic imaging can identify regions of interest for microscopic imaging.

The team decided that by initially looking at a single disease, they would be able to see where the gaps existed between different technologies. They agreed to use Alzheimer's disease as a model disease for the challenge, knowing that if they designed a good set of experiments, the procedure could be applied to other protein-folding diseases, and perhaps the challenge of integration in general as well

The Alzheimer's Disease Neuroimaging Initiative (ADNI-2) currently under way includes an extensive neuroimaging battery, but no EEG. The group believed that the existence of the program demonstrated the desire for more information on the effects of Alzheimer's disease on the brain, but that the study could be greatly improved upon.

Developing "Ideal" Experiments

Most microscopic imaging techniques cannot be used on living human subjects. The adult human skull is currently too thick for optical microscopy or photoacoustic tomography (PAT) to penetrate. The group concluded that their studies would have to begin with an animal model. Both animal and human model systems would need to be developed in order to match up cognitive degeneration with brain images.

The overall goal of the animal experiments would be to identify imaging correlates of cognitive dysfunction and progression. Because several transgenic mouse models already exist for Alzheimer's disease, the team would select one of those animals for use in its experiment. They would monitor cognitive impairment in the animal and conduct a battery of macroscale imaging techniques, including positron emission tomography (PET) to determine the time course of plaque formation and metabolic change, PAT of the hemodynamics, diffusion tensor imaging (DTI), EEG, and MRI. At the same time, they would conduct in vivo microscopic imaging experiments, including dual-labeled PET/PAT of different stages of protein aggregation. As PAT is capable of both microscopic and macroscopic imaging based on the same contrast, it has the potential to bridge the gap between images acquired at vastly different length scales. At different stages of cognitive impairment, some of the study animals would be used for ex vivo and post mortem microscopic imaging to determine the intracellular and extracellular localization of aggregates and to confirm the pathology via identification of plaques and tangles. Finally, the team would conduct proteomics experiments. The data from invasive or post mortem microscopy techniques in animals could be integrated with the data from the macroscopic techniques and help improve those non-invasive techniques so that, when the non-invasive techniques are used in humans, researchers can extract more information from them.

After the experiments with the mice were finished, the researchers would move on to human subjects, following many of the same procedures, but using the results from their earlier work to limit the number of imaging techniques used on the human subjects. PAT has not been used in humans before, so experiments on the brains of infants and the retinas of adults may be necessary before the technique would be useful in studying an adult brain. However, the thinness of the adult human cribriform plate could permit direct physiological measures at both microscopic and macroscopic scale in a deep cortical structure for the first time. These determinations would provide "ground truth" measures that can serve to meaningfully integrate across other imaging methods. The cribriform plate lies just below the orbital frontal lobe, which modulates reward and punishment processes.

After the imaging battery was completed, the team would be able to confirm the imaging correlates of cognitive dysfunction and disease progression. If the experiment led to unexpected findings, the data then would feed back into the animal model for further investigation. Finally, the procedure would require statistical methods for multiscale integration of high-dimensional data confirmation.

Technical Hurdles

At one point, the group made a wishlist of all the technological features they wanted on the imaging modalities they currently use. They wanted a way to measure bioelectricity at high resolution, GPS-style scalability, with which they could use landmarks to identify an area studied in a microscale technique and also study it with a macroscale technique (or vice versa). Finally, they wanted a dye sensitive to depolarization in neural cells, which would allow for imaging of the early signature of the disease. The dye would be particularly important, because it would be necessary for the PAT/PET experiments, the crucial link between microscale and macroscale data.

Concluding Remarks

Neuroimaging is expensive, and even while creating a wishlist of new technologies and talking about developing extensive batteries of tests for early disease detection, the team suggested that one long-term goal of the project should be to reduce the amount of imaging needed to diagnose Alzheimer's disease.

In their concluding presentation, the team remarked on the need for a "two way street" between microscopic techniques and macroscopic techniques. "It's a cycle of going back and forth, which we think is a solution," the presenter said. "When you're using one technique at one scale, you have to have the other techniques in mind." The ability to work with multiple techniques will help researchers compare and contrast imaging data by concurrently collecting datasets in animal models and humans.

IDR TEAM MEMBERS—GROUP B

- Chandrajit L. Bajaj, University of Texas at Austin
- Robert J. Barretto, Columbia University
- Graham P. Collins, Freelance Science Writer/Editor
- Richard S. Conroy, National Institutes of Health
- Scott T. Grafton, Director, University of California, Santa Barbara
- Daniel P. Holschneider, University of Southern California
- Andreas Jeromin, Banyan Biomarkers, Inc.
- Allen W. Song, Duke University Medical Center
- Kamil Ugurbil (IOM), University of Minnesota
- Gordon X. Wang, Stanford University
- Keith Rozendal, University of California, Santa Cruz

IDR TEAM SUMMARY—GROUP B

Keith Rozendal, NAKFI Science Writing Scholar,
University of California, Santa Cruz

In 1990, President George H.W. Bush proclaimed the decade beginning January 1, 1990, to be the Decade of the Brain, pointing to "advances in brain imaging devices . . . giving physicians and scientists ever greater insight." Twenty years later, further advances in neuroimagery continue to emerge at a rapidly accelerating rate, producing new challenges to realizing the benefits of brain research.

Neuroimaging techniques capture detail at sizes ranging from the atomic to the whole brain. Beyond the views produced by methods keyed to specific size scales, different imagery methods also track the nervous system over different time scales—from mere milliseconds to measurements taken across minutes-long experimental tasks or development courses that can span much of the lifetime of an organism.

Humpty Dumpty Has Fallen

As each method develops its own technology, training, literature, and theoretical paradigm, a real danger of fragmentation emerges. A global, comprehensive, and integrative perspective on the brain and nervous system may be more difficult to produce as more new imaging techniques emerge. A flourishing of new methods and technologies providing distinct insight on neural systems produced this situation. But it is hoped that the technological and methodological ferment may also hold keys to developing a coherent brain science.

The challenge before neuroimaging can be addressed by locating points on the horizon where the possibility of integration dawns. An Interdisciplinary Research team (IDR 1B) tackled this challenge during the 2010 National Academies Keck *Futures Initiative* Conference on Imaging Science. Their discussion of integration strategies followed a series of key questions posed by the steering committee that shaped the conference agenda and assembled the teams.

An interesting provocation at the beginning of this team's work helped to spark some creative tension that drove much of the early discussion. With a grin and perhaps a wink, one team member introduced himself as a serious skeptic of neuroimaging's value. This group member asked: "For all of the government and private foundation investment in new neuroimaging technologies and studies—perhaps hundreds of millions of dollars—what has that bought society?" He argued that such a major research initiative should have long ago produced abundant evidence that it promotes quality of life, medical successes, and other broad social benefits. Such a skeptical perspective would therefore add the corollary "Why?" to each of the questions posed to this IDR team.

Can All the King's Horses and Men Put It Together Again?

The team recognized aspects of the ancient debate on reductionism within the first of these challenge questions: "Can we establish a common language that unifies the data across all of the different levels of neuroimaging?"

Scientists have long recognized that reductionism, a powerful means of analysis, produces trade-offs with systems-level understandings. In the most successful cases, one can start from fundamental physical processes, like the kinetic energy of atoms in a gas, and fully reconcile this model with a larger scale model or measurement like air temperature, and beyond that to local air pressure, microclimate models, and on up. Could neuroimaging data be used to similarly integrate our understanding of the brain from the bottom up?

This would require the integration of models explaining ion channel processes, neurotransmitter actions, single neuron-biology, -genetics, and -signaling. These units in turn compose circuits and networks of neurons, cortical column organization, and regions and lobes of the brain. The Human Connectome Initiative aims to connect every brain cell to the others, a comprehensive map of all of the potential circuits in the brain. Combined with functional data, the active circuits within these connections could be determined. Because the team included people studying the brain at a wide range of scales, proposals for top-down approaches and questions about the wisdom of pursuing bottom-up integration repeatedly emerged.

The framework the team adopted assumed that the ultimate goal was to put the pieces together again, but how? And with what tools? And, of course, Why? Shouldn't the effort produce new discoveries, critical studies settling debates in the field, and the like? The team wanted to get more out of an integrated approach to neuroimaging data than could be produced by retaining the fragmented status quo.

Seeking Out the Right Glue

A recurring discussion point emerged concerning whether the integration should be a structural description of the brain or instead a functional or computational model? And wouldn't each type of model, within each strata of detail, and also globally, need to integrate and constrain the others?

Team members repeatedly related these questions to the need for a "gold standard" or fundamental element of the brain around which integration can be built. The team insisted that this gold standard needs to incorporate both structure and functional aspects. Some of the fundamental units proposed included the electrophysiology of a signaling neuron, the connections and neurochemical specialization of receptors and nerve cells, small circuits of neurons connected together, or the mini-columns found to be core structures organizing the cortex and often serving specific functions.

1. A physical glue?

The team raised an issue with an assumption within their challenge questions: that the ultimate "common language" needed to unify the diverse data and subfields in neuroimaging will be computational. Applications of mathematics to this challenge did attract significant discussion, but the team spent some time discussing a perspective that instead sought out a physical property that could tie together the diverse methods of neuroimaging.

The specific physical indicator representing often very diverse leverage points revealing distinct processes, can produce divisive forces. The team sought out a signal that could be used for neuroimagery across wide time and space scales. Progress on this front would facilitate integrating datasets because it would maximize the overlapping physical processes across the imaging modalities. An example of the difficulties that arise when linking incompatible signals can be seen in efforts to relate the BOLD signal of fMRI, primarily revealing metabolic processes, to neural signals, produced by electrochemical processes.

The team proposed focusing on using the electromagnetic fields produced within and between neurons as a unifying physical process to bridge the strata of measurement. There are static (field potential) and dynamic measures (spikes or EEG) of neural signaling at nearly every level of space and time resolution. The electromagnetic character of activities from the atomic to the tissue level should by necessity relate to one another according to well-understood physical laws. And this should help the integration process.

However, the team was wary of being seduced by the fact that current neuroimaging methods are heavily biased toward detecting signals in the electromagnetic spectrum. The historical success of electrophysiology methods in neuroscience may have led to this bias. Non-electrical physical processes also may hold some promise as a standard evident at every level of neural function. Some of these strata-spanning methods could be focused on the dynamics of chemicals within and between neurons or genetic inhibition and expression.

2. A computational glue?

Regardless of the physics of the signal, the team also pursued a potential common computational approach for mapping and integrating neuroimaging data between the different scales.

Here the team focused on the future promise of applying graph theory and other means of representing data in a common framework, abstracted from the underlying physical reality. Once neuroimaging data can be represented in the language of nodes and links, connections between levels of space or time become mathematically tractable. For instance, a graph-based model of several neural circuits could be used hierarchically with a higher level graph representing networks of circuits in a small volume of the brain. The lower-level model serves as an input influencing the state of a single node of the higher-level model. In this way, if all of the links between layers can be determined, the comprehensive model will unify the spatial levels.

Such a unification could be useful for what the mapping functions can tell scientists about how smaller-scale processes produce effects at a larger level, and how feedback flows down the levels to influence the more microscopic processes.

Other advantages that the team discussed for this approach were that these mathematical models could be produced directly from data or validated with real neuroimaging data and that the models easily incorporate dynamic or time-based variables, which better model the ever-active brain. Calculating correlations observed between real imaging data that bridge levels of time and space in this manner will help identify some of the coherence in the nervous system's structural and functional organization.

Thus, the abstract representation of information and its flow that graph theory produces could serve as a gold standard unit that helps align data from different levels of brain imaging.

Reassembling the Puzzle, Seeking Pieces That Fit

The team adopted an ambitious goal in seeking to unify all the layers of neuroimaging in a common modeling approach, but in the end recommended less ambitious sub-goals. Low-hanging fruit remain in the orchard of neuroimaging techniques awaiting integration. The team tried to identify areas where the integration between space or time strata seemed most promising in the near future. Out of some of these small-scale bridging successes, some general strategies useful for the other gaps could emerge. The team suggested that neuroimaging scientists should seek out ways of incorporating data at one level above and one level below their current preferred neuroimaging tool. These nearby methods should be most likely to give them insight into their current research questions.

Absent a physics-based gold standard that can simultaneously signal both structural and functional aspects of the brain, can another pathway be pursued to best produce integration? Simultaneous measurements at roughly the same spatial scale, using pairings or tripling of methods could help integrate across time scales as well as bridge the structure and function dichotomy. Here the discussion usually proposed solutions or discussed new developments related to combining a structural neuroimaging method like MRI with a functional method like EEG.

Several areas of fruitful convergence across one or two scales of time or space were discussed, including simultaneous EEG+MEG studies, fMRI and electrophysiology studies, studies of local field potentials as they relate to the BOLD signal used in fMRI, calcium fluorescence microscopy plus electrophysiological measures, and two-photon calcium imaging combined with MRI.

Which Pieces Should Be Picked Up First?

Many of the neuroimaging techniques available require sacrificing the research subject, which obviously precludes all but the post mortem study of human beings. The team nevertheless wanted to push the limits of non-invasive techniques in order to use human subjects whenever practicable. An integration of the data between human and animal studies should be kept in mind, however. The choice of an animal model for the work demanding invasive techniques should be made for maximum compatibility with the research focus in humans. That research focus, moreover, should allow for imaging with as many methods as possible across the scales to be integrated.

1. Picking up the right brain building blocks

The team's discussion of the most promising target systems for study in humans and nonhuman animals focused on smaller-scale neural systems, completely mapped and understood in terms of predictable outputs from known inputs. These would best support computational language development and testing. Sensory systems such as visual cortex or the retina or olfaction could fit the bill here. Many motor systems have the same detailed understanding already in place. Sensory systems also recommend themselves because previous studies have shown that they are organized both structurally and functionally in particular patterns, such as columns or bands of similar cells activated by similar stimuli. These may in fact prove to be organizational motifs in the nervous system replicated in other areas like the hippocampus. An integrated understanding of such a potential "building block," and confirming its generality, could make rapid progress possible in other brain regions and systems.

The team shared a general consensus that the cortical columns level of detail represents a particularly "sweet spot" to target with multiple methods, mathematical modeling of the unit, and testing against empirical data. A fully integrated understanding of a cortical column as a target was described as "reachable." Columns lie at the middle level of spatial scale, and models linking the column to smaller-scale structures seem to be on the verge of development. Functional MRI imaging can resolve detail at the level of the cortical column now, and the volume of a column is not unthinkably large for higher detail structural and functional mapping using existing methods. With the connections mapped completely within a column, some electrophysiological models will apply that will integrate spike train data and produce predictions about the overall electrical signal that may be detected above the column by EEG, for example.

2. Picking up the right animal model

The team then discussed studies of animals aimed at supporting a building block effort. Team members discussed the advantages of using animals that are traditionally studied within neuroscience, such as the worm *C. elegans*. This simple animal's nervous system has been comprehensively mapped using electron microscopy, which produced a synapse-level connectome. Furthermore, the functional operations of the worm's connected neurons, usually referred to as circuits, are also known. However, the circuit, which seems to work as a coherent unit or building block in the worm may be different than the elements out of which human brains are assembled.

Zebrafish embryos, easy to study because of their transparency and rapid reproduction, are also a promising organism to study. Their nervous system has been studied in a way that helps us understand narcolepsy in humans. The gene disrupted in this disease regulates the development of a 10-neuron circuit that has been completely mapped with two-proton calcium microscopy. This reveals the circuit at the level of every connection, both internally and externally, supporting the modeling of inputs and outputs.

The team emphasized the necessity for checking emerging integrative models against empirical data which would allow for hypothesis-driven experiments to further validate the emerging model. This would be the final goal of model testing—seeking to move from correlation-dependent models to those that can successfully predict the outcomes of studies designed for high internal validity.

Here, the team saw great utility in conducting perturbation-driven experiments in real tissue, comparing the observed effects of lesions, transcranial magnetic stimulation, optogenetic methods, and other means of selectively disabling key elements of the system. Parallel perturbations in the abstract integrative math model would also be pursued to validate the complete model. Such approaches are already widely used to test both structural and functional models in a wide variety of circumstances. These investigators are currently driving the development of perturbation technologies that are compatible with existing functional imaging modalities like MRI.

Plastic or fiber optic instruments that can cause temporary disruptions while functional data are being collected via fMRI could help to drive great strides in integrative research.

3. What pieces are missing?

The team also spent some time "pushing the creative envelope," seeking glimpses of blue sky technologies and creative methods and having fun with the challenge. Still, they hoped to identify desirable new developments, still somewhat miraculous at this point. Foremost were technologies for neuroimaging that allow for imaging of natural behavior in awake animals. What possibilities exist for portable neuroimaging technology, either through miniaturizing existing technology or developing new means? Carbon nanotubes can be fashioned into highly portable recording electrodes that can be fixed into place, and some hope exists for building large field MRI and making portable only the necessary other elements like the field coil for functional imagery. Other suggestions, based on currently emerging research, suggest immobilizing animals but allowing them to navigate and receive feedback from virtual reality systems and microelectrodes implanted in key sensory and motor nerves. A current mouse spherical treadmill and toroidal display system was discussed as well as a system for studying head-fixed zebrafish "swimming" in response to false visual feedback. Finally, tracking neurochemicals and brain metabolic processes in real time could prove quite useful to the teams challenge. Microdialysis performed in a helmet that a rat could wear might be one way to achieve this goal.

Recruiting Women and Men to the King's Army

The roles of scientific institutions must be addressed in meeting the challenge. What changes to science education and training should be implemented? What organizational, financial, and institutional developments will best serve progress toward integration?

1. Medics needed

The discussion here focused on clinical applications of neuroimaging technology, perhaps seeking more of an answer to the "why" challenge, and less to the "how." It was generally agreed that the integration of neuroimaging techniques could produce important gains in medicine, but it was noted that the costs of neuroimaging block its adoption in the clinic, even of single modality imagery. PET and CT have a higher penetration rate in community hospitals, to the detriment of MRI usage. In this case, the better method is not adopted, a fact attributed to differences in the reimbursement rate and the costs of the equipment itself. However, the costs of MRI are falling. Neurosurgeons still primarily rely upon electrodes and cortical stimulation mapping when operating on the brain, where fMRI may be applied. Only in the placement of deep brain stimulation implants is this imaging technology preferably used. This may not be entirely due to costs; the technique first used and more widely known and taught has the advantages. To encourage the development and use of new imaging techniques, including future integrative technologies, the team wanted to focus on easing the dissemination of the new technology by lowering costs and by increasing the ease at which the new technology is adopted. This should influence the design of the technology being offered to clinicians and the availability of training in medical schools and hospitals.

2. Scientists needed

The team believed that adoption of new integrated methods by cognitive scientists will best be fostered by incentive-based approaches. That is, scientists will be inspired by other scientists who have already adopted more complete modeling and imaging approaches, achieved breakthroughs, and attracted funding. The cognitive science field quickly adopted imaging techniques after applications to cases in the field emerged. For example, fMRI's impact on the field was significant once its successes were published and further studies were funded.

Early adopters of integrated methods could be recruited into training efforts, but the challenges securing funding for training and bringing together scientists with diverse backrounds remain. The team suggested introducing single methodology experts to one another at interdisciplinary conferences, targeting those pairs or triplets of technologies that show the most promise of integration. At this point, the work is a long way from creating a common language or model that integrates across all scales, but bridges and fusions across two or three levels are possible. Besides the salient example of this year's Keck *Futures Initiative* Conference itself, the team noted that meetings and collaborations like those enivisioned by the team, are already occurring. As a research problem exhausts the utility of one image modality, investigators are spontaneously seeking out other methods at different time or space scales. The comprehensive modeling approach to neuroimaging should encourage and spur additional similar such activities, the team concluded.

IDR Team Summary 2

Identify the mathematical and computational tools that are needed to bring recent insights from theoretical image science and rigorous methods of task-based assessment of image quality into routine use in all areas of imaging.

CHALLENGE SUMMARY

There is an emerging consensus in the biomedical-imaging community that image quality must be defined and quantified in terms of the performance of specific observers on specific tasks of medical or scientific interest. Generically, the tasks can be classification of the objects being imaged, estimation of object parameters, or a combination of both. The means by which the task is performed is called the observer, a term that can refer to a human, some ad hoc computer algorithm, the ideal Bayesian observer who gets the best possible task performance, and various linear approximations to the ideal observer. For any task, observer, imaging system, and class of objects, a scalar figure of merit (FOM) can be defined by averaging the observer performance, either analytically or numerically, over many statistically independent image realizations. The FOM can then be used to compare and optimize imaging systems for the chosen task, but there are many mathematical and statistical details that must be observed in order to get meaningful FOMs in studies of this kind.

First, real objects are functions of several continuous variables (hence vectors in an infinite-dimensional Hilbert space), but digital images are sets of discrete numbers, which can be organized as finite-dimensional vectors. Imaging systems that map functions to finite vectors are called continuous-to-discrete (CD) mappings; a great deal is known about their properties if the systems are linear and nonrandom, but little has been done for nonlinear systems or ones that have unknown or randomly varying properties.

Because the FOMs are statistical, some collection or ensemble of objects must be considered, and stochastic models of the object ensemble are needed. For objects regarded as functions, important statistical descriptors include the mean object, various single-point and multipoint probability density functions (PDFs), the auto-covariance function, and the characteristic function (an infinite-dimensional counterpart of a characteristic function, from which all statistical properties of the object ensemble can be derived). Each of these descriptors has a finite-dimensional counterpart when the objects are modeled, for example, as a collection of voxels, but great care must be exercised in the discretization.

The object randomness leads to randomness in the images; therefore, it is important to understand how to transform the object statistics through the imaging system; again, nonlinear systems pose difficulties. In addition, there is always noise arising from the measurement process, for example Gaussian noise in the electronics or Poisson noise in photon-counting detectors. Statistical description of noise in raw image data for a specific object may be straightforward, but the resulting statistical descriptors must be averaged over the object ensemble in computing FOMs. Moreover, many imaging systems themselves must be described stochastically. When image processing or reconstruction algorithms are used, all statistical properties must be expressed after the processing, because that is the point where the observer performance is determined. Interesting algorithms are often nonlinear.

Two issues are mentioned: how one quantifies the uncertainty in estimates of FOMs and how one determines the statistical significance of differences in estimated FOMs. Approaches to this issue include bootstrap and jackknife re-sampling, Monte Carlo simulation, and theoretical analysis.

All of this requires efficient and realistic simulation tools. The objects, systems, processing algorithms, and observers must all be included in a complete simulation, and the code must be validated.

Key Questions

- What current imaging applications would benefit from applying the principles of task-based assessment of image quality? What are the current methods of image-quality assessment in each? What are the important tasks? Are human observers customarily used?
- What new applications would open up in various fields if we used

higher-dimensional images, such as video sequences, temporally resolved 3-D images, or spectral images?

• For each application identified above, how should one model the imaging system? Are nonlinear models needed? Should the systems be described stochastically? What simulation code is now used?

• Again for each application identified above, what is known about statistical descriptions of the objects being imaged and of the resulting images? What are the important noise sources? Are statistical models used currently in image analysis or pattern recognition for this field? Are databases of sample images readily available?

• What new mathematical or computational tools might be needed for the applications identified? Are new image reconstruction algorithms, or new ways of applying and analyzing existing algorithms, needed? Is further work needed on noise characterization, especially in processed or reconstructed images? Are current simulation methods fast enough and sufficiently accurate?

• Is it important to have assessment methods that use real rather than simulated data? What gold standards would be used for assessing task performance with real data? Would there be interest in methods for assessment with real data but with no reliable gold standard?

Reading

Barrett HH and Myers KJ. Foundations of Image Science; John Wiley and Sons; Hoboken, NJ, 2004.

Barrett HH and Myers KJ. Statistical characterization of radiological images: basic principles and recent progress. *Proc SPIE* 2007;6510:651002. Accessed online June 15, 2010.

Clarkson E, Kupinski MA, and Barrett HH. Transformation of characteristic functionals through imaging systems. *Opt Express* 2002;10(13):536-39. Accessed online June 15, 2010.

Kupinski MA, Hoppin JW, Clarkson E, and Barrett HH. Ideal-observer computation in medical imaging with use of Markov-chain Monte Carlo. *J Opt Soc Am A* 2003;20:430-8. Accessed online June 15, 2010.

Kupinski MA, Clarkson E, Hoppin JW, Chen L, and Barrett HH. Experimental determination of object statistics from noisy images. *J Opt Soc Am A* 2003;3:421–9. Accessed online June 15, 2010.

Because of the popularity of this topic, two groups explored this subject. Please be sure to review the second write-up, which immediately follows this one.

IDR TEAM MEMBERS—GROUP A

- Alireza Entezari, University of Florida
- Joyce E. Farrell, Stanford University
- James A. Ferwerda, Rochester Institute of Technology
- Alyssa A. Goodman, Harvard University
- Farzad Kamalabadi, University of Illinois
- Matthew A. Kupinski, University of Arizona
- Zhi-Pei Liang, University of Illinois at Urbana-Champaign
- Patrick J. Wolfe, Harvard University
- Michael Glenn Easter, New York University

IDR TEAM SUMMARY—GROUP A

*Michael Glenn Easter, NAKFI Science Writing Scholar,
New York University*

In November, the National Academies Keck *Futures Initiative* brought together top imaging scientists from across the country for an interdisciplinary conference on Imaging Science. IDR Team 2A was asked to consider the mathematical and computational tools that are needed to bring recent insights from theoretical image science and rigorous methods of task-based assessment of image quality into routine use in all areas of imaging.

Team 2A was comprised of researchers armed with a broad arsenal of imaging knowledge, including expertise in consumer imaging, which is imaging that deals with products for consumers, optical imaging, and imaging in engineering, astronomy, and computer science. During two days at the conference, the IDR team debated long and hard about the best way to identify the tools that are needed to bring insights from theoretical image science and rigorous methods of task-based assessment of image quality into routine use in all areas of imaging.

These images could be of anything: a tumor, land that has been burned by a forest fire, or a prototype of a part for an automobile. Unfortunately, no image is perfect. It is likely there will always be errors, but the discussion of Team 2A aimed to illuminate these errors so that uncertainty in images could be minimized—and for good reason.

Imagine, for example, that you are a doctor. One day a patient is referred to you who is exhibiting signs of a brain tumor: impaired judgment, memory loss, and impaired senses of smell and vision. All of the signs are there. You run an MRI scan of the patient's brain. Once the scans come back, you scrutinize them. What do you find?

Because you know what you're looking for in the image, what you see may ultimately depend on the accuracy and detail of the image. Inaccuracy and insufficient detail are imaging's enemy; they make an image less true and therefore less useful than it needs to be. If you, as the doctor, can estimate and compensate for imaging errors more accurately, the image becomes more useful for the *task* of providing better patient care.

The task is a critical part of this scenario. Above, you were probably looking for a tumor, so the "task" of the image from the MRI scan was to show the presence or absence of a tumor. Any errors in that image are thus made more or less relevant depending on whether the error affects your ability to see a tumor in that image. This is the essence of task-based assessment of image quality.

In task-based assessment, the "quality" of the image is determined by its usefulness to the scientist, doctor, or other professional using it (the "observer"). This usefulness can be quantified, and it often needs to be if the observer wants to know how helpful the image is going to be or how good an imaging system is. This quality score for the image can be termed a "figure of merit" (FOM).

An FOM can be any measure of the image's quality. In task-based assessment of image quality, however, the FOM should ideally represent the ability of the image to help the observer complete the "task," whether the task is detecting a tumor on an MRI, measuring the power spectrum of microwave background in astronomy research, or classifying a forest as deciduous versus coniferous from a remote sensing image.

For an imaging system, the FOM represents performance ability, that is, how helpful is the produced image. System performance is impacted by many factors, error perhaps being the most significant. With that in mind, the team began to decipher, discipline by discipline, how to identify distinct sources of error in imaging. Once these identifications could be made, then their effects could be evaluated. As the team began to see the similarities between the sources of error in various fields, they began to reevaluate the textbook definition of the imaging process.

Traditionally, the imaging process includes: (1) the object, which is captured by the (2) imaging system, at which point (3) noise is introduced before the image is viewed by the (4) observer, who then can assign the image an FOM.

But this framework does not account for all the sources of uncertainty that affect the performance of a system and the ability of an image to aid in task completion. As the team discussed the many steps during the imaging process into which uncertainty could creep, a pattern emerged that prompted a new-and-improved flowchart of events in the imaging process:

1. The object is <u>illuminated</u> by a passive or active source, during which uncertainty exists in the illumination's spectrum, intensity, direction, and time (as well as interactions between those variables).

2. The <u>object</u> itself, whether it is real, phantom, or simulated, has uncertainty in its physical and biological properties.

3. The <u>emergent radiation</u> that will be captured by the imaging system has spectral, temporal, and spatial variation that can introduce error. Emergent radiation from a number of sources around the object can distort the image at this step.

4. The <u>imaging system</u> has multiple sources of error and uncertainty, many of which are specific to the imaging modality and field of study, that include management of noise and instrument calibration.

5. The system generates data that must be <u>processed</u> into an output image for the observer. This often involves reconstruction algorithms, general restoration (including noise reduction), and specific processing geared toward the specific observer. These processing steps may introduce information loss, artifact generation, or other error.

6. The <u>observer</u> now views the image. The observer can be human (using visual and cognitive systems to interpret the image information), algorithmic, or a combination of the two, and different observers will have varying levels of experience or training—all additional sources of uncertainty that can affect the image's usefulness and thus the assigned FOM.

Once arrived at the point in time to judge the image—to determine the FOM—we have encountered errors at every above step in the image's creation, which lead to a less perfect image, at each and every step, the final step being the sum of those errors. How each of these errors affects an observer's ability to use the image in a given task is specific to the task—a single image may be given different FOMs by different observers performing different tasks.

Here's the catch: If an imaging system is used by multiple observers with multiple tasks, optimizing system performance using a task-based method may not help all observers (and thus all tasks) equally. Similarly, general strategies to improve imaging modalities will not necessarily be relevant across fields. Identifying *which sources of uncertainty* negatively affect the FOM for a given task is the critical step to improving system performance.

All of the steps in which imaging errors occur build upon themselves, making a less perfect image—but an imperfect image may still allow the observer to complete the task. Perfect images would be ideal, but optimally *performing* images (images with perfect FOMs) might be the more prudent goal. If the next step in this process is to reduce errors, the logical question is not only which parts of the process *can* I improve, but also which parts will make the *FOM* improve? If each source of uncertainty and error is a knob on a large control panel, which knob(s) do I tweak to get what I need?

The team's answer to this emerging question was a vision: a new, refined approach to imaging systems that, depending on the object being imaged and what needed to be gained from the image, various settings could be manipulated to reduce the most relevant sources of uncertainty for that task, like a control panel with various knobs available for tweaking. The system could thus allow a balancing act, shuffling the amount and type of errors to optimize the performance of the imaging system for each given task.

IDR TEAM MEMBERS—GROUP B

- Ali Bilgin, University of Arizona
- Mark A. Griswold, Case Western Reserve University
- Hamid Jafarkhani, University of California, Irvine
- Thrasyvoulos N. Pappas, Northwestern University
- P. Jonathon Phillips, National Institute of Standards and Technology
- Joshua W. Shaevitz, Princeton University
- Remy Tumbar, Cornell University
- Tom Vogt, University of South Carolina
- Emily White, Texas A&M

IDR TEAM SUMMARY—GROUP B

Emily White, NAKFI Science Writing Scholar, Texas A&M

Key Questions
(modified by IDR team from original assignment)

How can task-based assessment be achieved? What approaches, if any, are already being used?

How do we define a task, and how do we define a figure of merit (FOM) for that task? What aspects of the imaging chain should be considered in assessing task performance?

Can models be used to assess image system performance? Should we use them as such; if so, how? Are current simulation methods fast enough and sufficiently accurate to aid in performance assessment? Is it important to have assessment methods that use real rather than simulated data?

How do we put task-based assessment into practice? What are the potential challenges involved in implementation of assessment approaches?

Task-based Assessment

Ideally, task-based assessment of image-system performance would include all participants in the imaging "chain": input (the object), system (the data generation), and observer (human or algorithm). All aspects of this chain would be statistically described, and predictive models would be used to test the performance of images, assigning to each system a task-based figure of merit (FOM). These FOMs, which would often be multidimensional to capture the maximum information about system performance, would then be used to compare performance across imaging modalities, thus discovering which systems perform best at a given task. These theoretical models of the imaging chain would also be used to simulate image output from theoretical imaging systems in order to decide which hypothetical systems are most prudent to build and use.

For both simulations and real-life testing of a system, again in an ideal scheme, standardized inputs would exist to maximally inform the FOM. Databases of such standardized input ensembles, and of the gold standard output ensembles, would exist for all conceivable tasks. For non-simulated tests, easily transportable calibration samples would be validated at multiple locations and then used to evaluate new systems and any modifications to existing systems. Observer ensembles would also be used in assessing task-based performance to account for variation in user decision making, especially with human observers.

In generating the FOM, ideally, error assessment would account for the fact that not all errors are equal—unlike (and in this case possibly superior to) a receiver operating characteristic (ROC) curve, which represents all task failures as equivalent points on a curve. Grievous errors (for example, missing a large tumor in a dangerous location) would be ranked as more serious within the error assessment process. Conversely, easily identifiable errors (such as a completely jumbled image that clearly does not resemble a typical image) would be ranked as less serious by the error assessment because they are easily recognizable and would likely not lead to serious adverse outcomes. These weighted performance outputs would be task specific.

Practical Considerations

The scenario above is desirable but currently impractical. Challenges of course exist that will impede the development and implementation of such performance assessment approaches.

Generating standardized inputs

Theoretical models of imaging systems are currently not sufficient to inform performance assessment. Although current models might be helpful in the development of new systems, the statistical descriptions of system components are not currently complete enough for model simulations to be fully predictive of system output, necessitating assessment approaches that use real inputs and data. However, standardized input ensembles also do not exist, and we currently lack an understanding of how many and what variety of images would be needed to best inform assessment.

Even with standardized input ensembles and datasets, ensemble optimal performance does not guarantee optimal performance on individual images. FOMs may not describe system performance as it applies to extreme cases, and these cases might in fact be the most critical—the outliers that you truly need your system to perform well on. Therefore, it remains unclear how to use an FOM to optimize a system when average performance may not correlate with performance on critical inputs. Similarly, standard input ensembles need to take into account the varying impact of different errors. As discussed above, not all errors are equal, and input sets need to include sufficient variety to allow for detailed error analysis. Unanswered, unfortunately, is the question of how to decide which errors are high versus low impact, how to weight errors based on these impacts, and whether this procedure will induce even more bias into an already noisy system.

Most and perhaps all relevant fields also lack gold standard data, and for many systems it is impractical to generate such an ensemble because the ground truth is often unknown—for example, exploratory images, such as many imaging endeavors in astronomy, cannot be evaluated based on a "correct" answer, because the input has not yet been characterized. One problem with such images is processing; for example, removing noise from these images can be detrimental if the noise is relevant, but it is often difficult or impossible to know whether noise in these images is in fact a real and interesting phenomenon. Defining performance without knowing ground truth may be a prohibitive consideration in approaching task-based performance assessment, although some systems have robust imaging systems despite unknown inputs.

Evaluating the imaging system

Once a standardized input ensemble has been established, the next step is to determine which aspects of the imaging system are practical to consider when assessing performance. For example, the observer is likely a highly influential part of the imaging chain with respect to task performance, and uniform, reproducible performance by human observers may be unlikely. In many current imaging applications, a human is still the ideal observer given current technology. Especially for tasks like medical diagnoses, this is unlikely to change in the near future. Thus, in generating an FOM for use in optimizing a system, assessment should account for the costs of retraining the human observer. Systems should be optimized for the actual observer, not an abstract ideal observer—even in theoretical imaging models. Otherwise, an important aspect of the imaging chain is ignored.

For example, when an observer performs a task successfully or with errors, is it because of some component of the imaging device (as we would assume in optimization approaches that do not account for observer biases and preferences), or is it because of biases resulting from the observer's tacit knowledge? Or could performance even be affected by some subtle issue in the particular way we defined the task? Improving image quality can actually impede human observers trained on noisy data. The ability to feed information back into the system during optimization is thus hampered by the inaccurate assumption that an observer is acting in a reproducible way.

The expert human observer, nonetheless, is a critical, beneficial part of imaging systems because of the ability to incorporate tacit knowledge and real-time, non-image information into the task performance. However, it is impractical to use humans in the optimization process—the cost of human participation and the number of humans needed is likely prohibitive. It is also unlikely, however, that we can accurately model a human observer, because we lack the ability to account for the tacit knowledge and non-image information that exists in the human brain. Lack of ability to model good/realistic observers may thus also limit our ability to assess new techniques.

Thus, because a task exists within a larger framework, we need to optimize both modality and interpretation. We need better models to aid in optimization, and models need to account for the human observer in feeding back information. Metrics must also account for the fact that tacit knowledge complicates task-based performance assessment. Observer performance depends on more than ideal image representation, and human observers have preferences and limitations—and these may vary between even highly similar tasks. As such, it is possible that FOM-based assessment is an unrealistic approach for multiple-observer systems.

Moreover, an important aspect of the imaging chain is the translation of system data into an image output—a model of the data and thus of the object. For example, the raw data output of a modern 3-D ultrasound system would be virtually impossible for a person to visualize, yet data modeling permits real-time 3-D computer reconstructions that are easily interpreted by human observers. However, not all image outputs are ideal; some imaging systems have shortcomings in data modeling, thus limiting the capability of the image to describe what we want to know about the object. Other potential pitfalls are failures of the system to produce consistent images, the presence of artifacts in object reconstruction, or the inability to map aspects of an image to an object's physical characteristics. Having standardized inputs or accurate forward models (computer models of the image based on the object) is of little consequence in applications with poor (or poorly understood) data modeling for image outputs. It is thus important to understand and assess the data modeling within an imaging modality in order to understand its limitations and possible design improvements. Understanding data model limitations is especially important for ensuring consistent and reliable computer processing, which is often critical for high-throughput or large data volume applications.

Defining the task

Assuming the existence of standardized input ensembles and a well-defined understanding of the imaging chain, the question remains of how to define a task. Ideally, a task would map a scientific question to a system output. However, we can approach this question from the position that there exist two classes of imaging: a task-based class and an accuracy-based class. Because one cannot predict future tasks, and it may be useful to use one image for many tasks or have the option to use an image later in the performance of other tasks, maximizing object representation (improving the accuracy of mapping the object's physical characteristics) may be prudent. Object representation could be considered a task (the task could be defined as "maximum information gathering"), but this measure of performance is not traditionally considered task based. Nonetheless, object representation is important and relevant for the longevity and broad usefulness of an image (e.g., corroborating information from multiple imaging modalities, which was mentioned by other groups). Perhaps also important to note is that there may be scientific questions that involve image use that do not have a clearly identifiable "task." Task-based methods thus have relevant limitations.

An additional consideration in defining a given task is that one purpose of task-based assessment is to generate information that will aid in optimization. In this case, cost and usefulness must be considered (not just performance): For a system that performs many tasks, it may be more practical to simplify and broaden tasks for greater applicability. In other words, we could optimize a system for a range of tasks, even though this would likely lead to suboptimal performance on specific tasks. Alternatively, it may be possible to include various settings within a system that could be adjusted to optimize performance on individual tasks. The best approach to practical task-based optimization is thus unclear.

In defining an FOM, it is also prudent to consider cost issues. For example, small incremental improvements in system performance might be expensive to implement, so it is important to define what level of difference between two FOMs merits upgrades or system adjustments. These infrastructure considerations also should account for the observer (as discussed above): There is a cost to human learning, and retraining observers, particularly human observers, to perform tasks using an improved system output could be prohibitively costly. Also worth considering are cultural norms within the human observer community—in addition to the observer learning curve, the "newness" of an image or system could impede implementation because of observer rejection of optimized systems and outputs. These ideas should be considered in defining the FOM and in deciding how to evaluate significant differences between two FOMs.

For task-based optimization, other issues include practicality of modifying an existing system. Reduction of dimensionality (model parameterization) may be a prudent approach to optimization in order to simplify the process of system design or modification. Modeling would also be of great use for optimization procedures, but further noise characterization, as well as determination of how noise affects "object understanding," is necessary before modeling can be practically used. Another approach to system optimization is subset analysis— joint optimization of parts of the imaging chain or system. Adjusting components of systems in this way makes optimization tractable in otherwise highly complex systems, although this approach may not reach a global optimum even when one exists. To perform subset analysis, however, we would need to define subtasks, which also would require performance assessment. In attempting to optimize a complex system, a challenge will be to define tasks that address specific subsystems and maximize the chances that individual optimization will result in a global optimum.

Future Directions and Recommendations

Despite the challenges of designing and implementing image metrics of system performance, initial steps toward task-based assessment of performance are prudent and achievable. Individual fields can begin to identify and gather or develop standard input ensembles for widespread use. Scientists can also identify or develop gold standards for imaging and create databases, also for widespread use. These input and output ensembles could be used to assess existing imaging systems via a "round robin" approach (imaging one input ensemble using many systems).

Careful consideration can be given to understanding and assessing the data modeling of an imaging modality. Such assessment should be made with due consideration to the class of objects being evaluated and the related task. For example, optimal imaging of microscopic transparent samples likely requires a different imaging modality than searching and recognizing faces in an airport screening system. The study of the data model will reveal its limitations and thus help in establishing avenues of research for optimization. It is also worth considering whether the limitations of the data model are so prohibitive as to merit an alternative approach—for example, synthetic aperture radar captures data but no image, and that is sufficient and accurate for a representation of the object to be computed, recognized, detected, and classified (i.e., "algorithmically understood").

For analysis of system performance, one can devise a means of failure analysis: methods of ranking and weighting performance errors based on impact. We can begin to address potential methods of incorporating tacit knowledge into modeling and assessment. In doing this, we should also consider how we might adapt our approaches in the future to address higher-dimensional images and advanced imaging methods and also how to improve statistical descriptions of the imaging chain (including the observer) to achieve adequate modeling. It is also worthwhile to consider whether some simple models might have the capacity to adequately inform performance assessment.

One practical and achievable goal is to design observer-based systems: Imaging systems could include settings for personalized optimization guided by real-time feedback including personalized error scores. Different imaging protocols could be optimized for each observer using real-time calculation of different views, displays, and contrasts with adjustable parameters. These personalized imaging systems would thus rely on both observer-based and task-based assessment, perhaps more effectively addressing system non-idealities.

These approaches, of course, need to be examined by a wide range of diverse communities. Nonlinear systems, which have potential high impact on imaging capabilities, may benefit most from some initial steps toward task-based assessment because of their complexity and current lack of sufficient methods for assessment. Such fields include compressed sensing, deblurring or deconvolution, nonlocal means filtering and estimation, and spatiotemporal methods.

IDR Team Summary 3

Develop and validate new methods for detecting and classifying meaningful changes between two images taken at different times or within temporal sequences of images.

CHALLENGE SUMMARY

If a picture is worth a thousand words, multiple pictures of the same object are often worth a million. By comparing PET/CT images taken before and after chemotherapy or radiation therapy, a physician can often tell with high certainty whether a tumor is responding to the therapy. A military analyst looking at synthetic-aperture radar (SAR) images of an airfield can discern that a new type of plane has been deployed. A set of Landsat images taken weeks apart can be used to determine if a crop is flourishing or withering. An astronomer comparing serial images may discover a supernova or a gamma-ray burst.

Yet not all change is meaningful. Two digital images of the same object are never identical, on a pixel-by-pixel basis. The ambient lighting may change between aerial photographs; the patient might lose weight or lie on the scanner bed in a different position, or the crop images might be taken at different times after irrigation. Technical factors can also change: the magnification might be slightly different between two aerial images, or a different X-ray tube voltage or amount of contrast agent might have been used for two different CT images. These kinds of change are easily detected simply by subtracting two images, but the resulting difference image could still convey no meaningful information about the important changes for which the image are being compared. Focus is needed on change and how best to interpret change.

Current approaches to change detection are surveyed in the references below. Radke, for example, describes many sophisticated ways of normalizing images so that trivial changes in lighting or technical factors will not be called a change, and he introduces advanced concepts from statistical modeling and hypothesis testing; yet he stops short of application-specific "change understanding," his term for classifying changes as meaningful to the end user.

There is a strong need for developing rigorous methods not only for detecting changes between images but also for using them to extract meaningful information about the objects being imaged. One approach is the use of statistical decision theory, where the statistical properties of images of normally evolving spatiotemporal objects are modeled, and "meaningful" is defined in terms of deviations from these normal models. Alternatively, specific statistical models can also be devised for various classes of interesting changes, and in this case "meaningful" can be defined in terms of classification accuracy or costs assigned to misclassification.

A different approach is to recognize key components of the evolving images and their spatiotemporal relation to one another. This semantic approach is similar in spirit to what the human visual and cognitive system does in analyzing scenes containing well-delineated, temporally varying object components, but computer implementations can take into account the noise and resolution characteristics of the images.

For statistical or semantic approaches, or any synthesis of the two, there is a pressing need for assessing the efficacy of the change detection and analysis methods in terms of the specific task for which the images were produced. This assessment could then be used to optimize both the algorithms themselves and the imaging systems that acquire the spatiotemporal data.

KEY QUESTIONS

- What fields of application, within the expertise of the participants, require careful discrimination between meaningful and trivial changes? In each, what are the characteristics of meaningful change?
- In each field identified, what databases of imagery or other data can be used to build models of meaningful changes?
- Can fully autonomous computer algorithms compete with a human analyst looking for meaningful changes? How can the computer enhance the capabilities of the expert human? By analogy to computer-aided detection (CAD) or diagnosis (CADx) in medicine, can computer-aided change detection (CACD) be applied in the applications identified?

- What is the relative role of semantic analysis and statistical analysis in understanding changes?
- What modifications in the basic paradigm of task-based assessment of image quality are needed for tasks that involve temporal changes?

READING

Coppin P and Bauer M. Digital change detection in forest ecosystems with remote sensing imagery. *Remote Sens Rev* 1996;13:207-34. Accessed online June 15, 2010.

Radke RJ, Andra S, Al-Kofahi O, and Roysam B. Image change detection algorithms: A systematic survey. *IEEE Transactions on Image Processing* 2005;14(3):294-307. Accessed online June 15, 2010.

Singh A. Digital change detection using remotely sensed data (Review article). *Int J Remote Sensing* 1989;10(6):989-1003. Accessed online June 15, 2010.

Because of the popularity of this topic, three groups explored this subject. Please be sure to review the second and third write-ups, which immediately follow this one.

IDR TEAM MEMBERS—GROUP A

- Mark Bathe, Massachusetts Institute of Technology
- Felice C. Frankel, Harvard Medical School
- Ana Kasirer-Friede, University of California, San Diego
- K. J. Ray Liu, University of Maryland
- Joseph A. O'Sullivan, Washington University
- Robert B. Pless, Washington University
- Jerilyn A. Timlin, Sandia National Laboratories
- Derek K. Toomre, Yale University
- Paul S. Weiss, University of California, Los Angeles
- Jessika Walsten, University of Southern California

IDR TEAM SUMMARY—GROUP A

*Jessika Walsten, NAKFI Science Writing Scholar,
University of Southern California*

IDR team 3A wrestled with the problem of defining meaningful changes among images. These changes can be between two images or in a series of images over a period of time.

Analyzing images to detect changes from one to another is not as simple or straightforward as many of us think. Even when looking at two still images side by side it can be hard to differentiate what is meaningful from what is not. For example, two photos taken of the same section of forest at different times of day will have variations in light that may distort the meaningful changes, exaggerating or minimizing them. Some tools do exist, such as principal component analysis (PCA), that can help normalize the images, correcting for any background noise or variation. But the use of any analysis technique, whether it's PCA or model based, will vary depending on what the researcher is looking for. There is not a universal tool that can be applied across disciplines. Likewise, each researcher will run into different problems during the analysis of data from different imaging technologies. To illustrate, a biologist looking at vesicle fusion within a cell may run into issues with the image resolution produced by the instrument he or she uses or instrument vibrations. On the other hand, an analyst looking at color change in leaves may run into problems with the intensity of sunlight or wind.

Because there are so many variables at play when images are analyzed (e.g., instrumentation, light, vibration, resolution, etc.), the IDR team thought it necessary to somewhat narrow the scope of its original challenge, which was to: Develop and validate new methods for detecting and classifying meaningful changes between two images taken at different times or within temporal sequences of images. The group focused instead on two aspects of this statement, altering it to read as follows: Develop and validate new methods for detecting and classifying meaningful **trends** within **temporal sequences of images**.

From these temporal sequences, trends need to be detected from the data, not just changes from one image to another, so that researchers can model what is happening over time and then use those models to predict the outcomes of future experiments. These trends are more meaningful overall than just defining what changed between two images.

Exploring Terms

It's easy to get hung up on terms, but sometimes it is helpful and necessary to define terminology. In the case of the group's redefined statement, three ideas need further vetting.

First, the images that need analysis can come in a variety of forms. They can be still photos, a handful of snapshots from video surveillance cameras, or hundreds of hours of video. These images can be two-dimensional, three-dimensional, spectral, four-dimensional (three-dimensional plus spectral), or even five-dimensional (four-dimensional plus time).

With all of these different types of images, it can be complicated trying to assess them, especially when all of the variables are taken into account.

The second word that needs some explanation is meaning. What does it mean to be meaningful? The group determined there are two kinds of meaningful processes: exploratory and explanatory. Exploratory processes lead to discovery or surprise. In this case, a researcher may not go into an experiment knowing what he or she is looking for and is surprised by the finding. Explanatory processes are the analyses in which a researcher will attempt to make sense of data, reducing pages and pages of numbers to something meaningful.

Meaningfulness can be quantified in a number of ways. Specifically, the IDR team talked about information theory entropy measurements where entropy, a measure of randomness, is compared with the probability an event will occur. This relationship is inversely proportional. For example, if a vesicle fusion event is likely to occur many times during a short period of time, the chance that something random will happen (entropy) is much lower. Quantification of meaning can also occur through measurements of error in the data. Recognizing error can be difficult. Oftentimes, it involves reanalysis of the data using trial and error to find the information that is important to the experiment.

The word trend is similarly ambiguous, meaning different things to different people. In general, however, the team defined a trend as meaningful changes over time. Trends are evolving processes that have directionality. There is usually a growth and collapse phase in a trend, but a trend may not necessarily go in one direction (i.e., it can go up and down multiple times). By measuring a trend a researcher can say something more about the process, possibly using the trend as a predictive model.

Trends can be found in birth or death rates, morphology, structure, topology, particle motion, diffusion, flow, drift pattern, spectral, background, noise, intensivity, reflectivity, transmissivity, density, and statistics (non-visible).

Team 3A noted that trends can be broken down into categories. These categories include monotonic, linear, periodic, random walk, or impulsive/frequency. The data from an experiment usually doesn't nicely fit into one trend category. Rather, the data is a combination of multiple trend categories. Tools exist to decompose a temporal sequence of images into trends, but those tools only analyze data from one category. So, the challenge then is to find methods to decompose temporal sequences of images that have contributions from multiple categories of trends.

Also, some of these trends may be more or less meaningful than others, and there may be trends that compete within an application, distracting the researcher from what he or she is looking for. It can be difficult to extract the meaningful trends from the non-meaningful trends. For example, the patterns of vibration in a video of cell vesicle fusion are not related to the vesicle or even the cell. The vibrations come from the instrument used to capture the video. But background noise, like instrument vibrations, are not always as easy to detect.

How Do You Find Trends?

Both explanatory and exploratory processes are used in experiments to find trends. A researcher first goes into the exploratory phase. A scientist may go into the experiment knowing what he or she is looking for. But that is not necessarily the case. This exploratory phase leads to discovery, which then helps the researcher formulate or reformulate hypotheses. That can motivate the experiments. The researcher then moves into the explanatory phase to attempt to support or invalidate the hypotheses.

Researchers can use both factor analysis and dynamical models to analyze their results. Factor analysis looks at the raw numbers and finds trends in the numbers. For example, if a video of cell vesicle fusion events is analyzed via a statistical program, like MATLAB, the program will average all of the images in the video into a composite image. The researcher can then look at a specific part of the video, creating an image that represents that parameter. That video is then compared at certain intervals to the average image, and a graph that shows how far the video is at any point from the composite image is produced. This graph will show a trend in the information that can indicate a specific event, like vesicle fusion, did or did not occur. Depending on the parameters used, the trends produced may or may not be useful. So other methods could be employed to analyze the data, such as dynamic models that use time as a comparison to certain points.

Mathematical analysis can be used to find trends in intensity or frequency of events in a temporal series of images. In the vesicle video, a flash of light represents a vesicle fusion event, a cellular mechanism important for cell movement and the transportation of cellular material. These flashes, or events, occur in varying speeds and intensities. A researcher could analyze the flash intensities or speeds to see what the trends are. Are vesicles fusing more quickly at a certain point? More slowly? Why? The findings could then be used to predict what would happen in further cell vesicle fusion experiments, completing the feedback loop. The importance of these predictive models may not be as apparent in the vesicle fusion example. Nevertheless, cell biologists may find these types of trends meaningful in future research.

If these types of models are applied to tumor growth, for example, a researcher may be able to predict the behavior of a tumor for certain locations in the body. In addition, finding these trends may help researchers better understand ways to redesign experiments.

Future Areas of Development

Many challenges are encountered during the experimental process. These include massive datasets, limitations due to instrumentation, resolution in time, space, and spectrum, trends on multiple time scales, data in multiple dimensions, and representation and communication of results. Overcoming these challenges will help lead to future progress.

Team 3C sees these future developments as falling into three types. First, researchers could use tools in new ways, such as PCA analysis to preprocess a video. Tools could be used in post-processing and removing unwanted noise in an image. This could also mean using tools or theories in disparate fields to help analyze the data or solve problems collaboratively. One example of this is pattern theory, a mathematical theory that tries to explain changes in images using combinations of a few fundamental operations. Pattern theory does not account for series of images over time. Other limitations of the theory also need to be addressed to develop more mature and implementable versions of pattern theory that can be applied to scientific image analysis.

Second, nonlinear representations of data need to be developed. Current methods of factor analysis are linear, accounting poorly for motion. In scans that involve deformation by movement, such as distortions in images from PET scans from breathing or objects on surfaces that are being deformed, mathematical models need to be built to account for the deformations.

The third and final recommendation for development was the use of iterative feedback for prioritized development of mathematical tools, instrumentation, and experimental design. This means using experimental analysis to reassess the original problem. For example, new instruments could be developed based on what happens during a particular experiment, and trends that are found could be used for predictive models.

If researchers develop these areas, they will better be able to find meaningful trends in images and find ways to improve their data analysis from those images.

IDR TEAM MEMBERS—GROUP B

- Daniel F. Keefe, University of Minnesota
- Lincoln J. Lauhon, Northwestern University
- Mohammad H. Mahoor, University of Denver
- Giovanni Marchisio, DigitalGlobe
- Emmanuel G. Reynaud, University College Dublin
- James E. Rhoads, Arizona State University
- Bernice E. Rogowitz, University of Texas, Austin
- Demetri Terzopoulos, University of California, Los Angeles
- Rene Vidal, Johns Hopkins University
- Emily Ruppel, Massachusetts Institute of Technology

IDR TEAM SUMMARY—GROUP B

*Emily Ruppel, NAKFI Science Writing Scholar,
Massachusetts Institute of Technology*

Current Imaging Methods: Not Always a Clear Picture

Imagine walking into an empty room, turning on the light, and taking a picture of a chair. Now imagine walking into that same room a week later, not turning on the light, and taking a picture of the chair using the flash on your camera.

In the two images, the chair would look completely different. But how can we tell whether the chair or the imaging actually changed between picture one and picture two? Are the differences significant, or not?

Let's assume, for instance, that the chair did change (perhaps the room flooded and the wood warped slightly out of shape). If one wants to use those two pictures, the first taken before the flood, and the other taken after the flood, to track the flood's effect requires knowing what is also different about the conditions of the lighting, the camera apparatus, or perhaps the stability of the room, itself. Because there are many possible sources of interference, and they cannot be isolated, the problem has no easy solution.

Now imagine a more probable situation. Your skull is being imaged to look at a tumor. Over a period of weeks or years, your neurologist is trying to determine whether the tumor is changing in any significant way. It should be possible to know by looking carefully at fMRI, PET, and/or CT scans, the widely trusted tools of neuroscience, without having to resort to surgery. Is it that simple?

This is exactly the problem that IDR team 3B tackled at this year's National Academies Keck *Futures Initiative* Conference on Imaging Science. If anything about the calibration of the MRI device changes, if there is some methodological change between two brain imaging sessions, the resulting data could suggest changes that have nothing to do with whether the tumor is shrinking or growing or spreading.

Meaningful detection of change is not just a problem in neurology—scientists in many fields are eager to establish better methods for capturing and determining significant change based on highly reliable imaging. Forestry experts need to know whether satellite or aerial pictures can be trusted to accurately compare canopy images over time. Likewise, astronomers must use still pictures to observe ever-changing celestial phenomena. Oceanographers, military surveyors, even farmers could benefit from advances in imaging science.

Mining for Meaning in Images

By modeling human behavior with computer programming—that is, encoding a process or calculation that humans used to do by hand into something the computer can do for them—computer scientists improve productivity and free up time and minds for solving other problems. If the same solution could be applied to imaging science, it would be enormously helpful for those scientists whose complex pictures, videos, and visual models must be painstakingly deciphered for useful analysis. IDR team 3B sees a key opportunity for improvement as a sharpening of imaging language. For instance, most people think of an image as an array of pixels, but team 3B's definition includes radiographs, 3-D graphics, even nonvisual media like audition and haptics. By specifying the meaning of the words scientists use to identify key features in an image set, they increase their chances of successfully teaching others to identify meaningful change and develop computer systems unique to their problems.

In the introductory example, the obvious subject of interest is the chair, and the obscuring factors are lighting, noise, and equipment. But there are many fields of science in which the lighting, itself, could be the subject that needs measuring. If that were the case, shadows on the chair would provide a way to detect lighting change.

In as wide and varied a field as imaging science, the best computer programmer is unlikely to come up with one solution that addresses challenges of comparing visual data in medicine, astronomy, and environmental science. With no "fix-all" that, when reduced to computation, could help determine meaningful change in every situation, IDR team 3B focused on developing a model that scientists in their respective fields can use to solve their own imaging problems, using creative algorithms where necessary and/or possible.

The Man Machine

While the vertical flow of this model focuses on the relationship between humans and their equipment, the horizontal arrows point to the heart of the matter: the relationship between humans and how they can use representations to help a computer "see" their data. Without meaningful

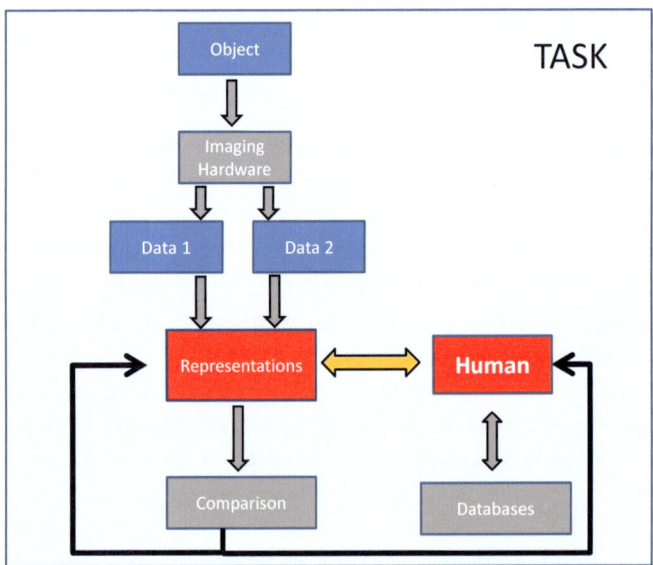

FIGURE 1.

representation, scientists continue to have to work from raw data, spending time deciphering what's most important in small and large datasets. By creating an algorithm that could do this step for them, and using the creativity inherent in human beings to continually redefine representation, scientists in many fields could save a lot of time.

The thin arrows that close the feedback loop from Comparison back to Human and Representation is part of the redefinition step. New methods will obviously require evaluation, and that evaluation will be used to tweak the representation in question.

The IDR team's model is dependent upon the human element in computer science—for instance, instead of letting the machine take over, you, the scientist, know your imaging problem best. You know what you want to see, and what you don't want to see. If humans become *active* participants in not only evaluating the effectiveness of their respective systems, but also reimagining the computers that are often merely their tools, they open up possibilities for creative computerization and previously unseen solutions.

The best example for how to value human creativity in this particular kind of problem solving comes from a rather surprising source: the unfolding of our own understanding.

Data Dreaming

One of the team members, years ago, had a dream. A dream about clouds. Although most scientists would not consider the whirring of one's subconscious a proper tool for problem solving, this particular dream cleared the air, if you will, of a seemingly impenetrable problem.

The team member needed to count clouds. The satellite images he was working with needed to be evaluated for clarity. For instance: was the milky character of the picture due to cloud cover, or was it snow on the ground?

In his dream, the team member saw the clouds in a paralax effect. (Paralax refers to how movement alters your perception of your surroundings. For instance, when you drive a car, the things that are close to you shift out of your vision at a different rate than things in the far background.)

Because the satellite was not just taking one picture, but five quick-succession snapshots, the dreamer realized that instead of looking at the combined images, cleaving those snapshots from one another and comparing the edges of clouds as the satellite moved was one way to determine how many clouds there were.

By applying this idea to his algorithm, he solved the puzzle in a creative way, and was able to turn his attention to other things.

Team 3B thus encourages continual participation by the human in the model, because by abandoning the computer to its work, a scientist will also abandon the possibility for improving and increasing the volume of work a computer can do. Research into the methods that humans use to understand data, and improving the relationship of all research fields with computer scientists in several disciplines will be a big step toward improvements in imaging science.

Although such a vision is not an entirely new idea (its been a big move in the computer science world since around 2005), IDR team 3B thinks that its introduction to and integration of all methods of imaging will help set the foundation to "begin building a new generation in human/computer interaction, which will enable us to envision a new era of understanding in the representation and analysis of complex images."

IDR TEAM MEMBERS—GROUP C

- Sima Bagheri, New Jersey Institute of Technology
- David A. Fike, Washington University
- Douglas P. Finkbeiner, Harvard University
- Eric Gilleland, National Center for Atmospheric Research
- David M. Hondula, The University of Virginia
- Jonathan J. Makela, University of Illinois at Urbana-Champaign
- Mahta Moghaddam, The University of Michigan
- Naoki Saito, University of California, Davis
- Curtis Woodcock, Boston University
- Olga Khazan, University of Southern California

IDR TEAM SUMMARY—GROUP C

Olga Khazan, NAKFI Science Writing Scholar,
University of Southern California

Scientists who rely on images to provide data are faced with an unusual challenge: Although taking two images is easy, finding the scientific difference between the two images remains a much more complicated task.

Scientists trying to find the differences between two images often find themselves constrained by the limited number of tools for image differentiation in their field. Someone who studies changes in the ocean floor, for example, would use a different methodology than someone who looks for changes in the urban climate. The type of software, the type of algorithm used to read the data, and even what is considered "noise" (or irrelevant information) are specific to each field, and they vary from discipline to discipline. Because of this segregation of image detection methods between physicists and climatologists, for example, researchers frequently find themselves "stuck" doing change detection as it has always been done in their field, which can stymie the progress of change detection methods overall.

IDR team 3C, at the National Academies Keck *Futures Initiative* Conference on Imaging Science, was tasked with "developing and validating new methods for detecting and classifying meaningful changes between images." Although standard methods for differentiating images already exist for everyone from astronomers to zoologists, scientists from different areas suffer from a lack of communication about these methods.

In order to help scientists get a more complete impression of the changes that occur between two images, team 3C set out to breach the divides between disciplines when it comes to image processing. The group aimed to lay a unified framework that would combine the practices used by everyone from astronomers to climatologists to radiologists.

What Is Image Differentiation?

There are three categories of observations a scientist might note when evaluating the changes between two images: First, there is everything in the image that the researcher is not interested in measuring, like rocks when the study is about trees, or stars when the study is about planets. There is also the noise/artifact, or the interference from environmental factors, such as soil moisture and cloud cover. Finally, there's everything the researcher is interested in measuring, which can also be called the meaningful change that occurred between the time two images were taken.

The easiest way to define a meaningful change in an image might be simply "a cluster of points in a large space" as one of the group's researchers explained.

That meaningful change usually has a few defining characteristics. First, it is persistent, in that it appears repeatedly throughout multiple images. Second, it is specific to a certain portion of the image. That is, a large portion of the image will remain the same, but the change occurs in a small area.

Typically, the way differences between two images are found is through the following chain of actions: First, the scientist captures the images. Then, the images are examined in order to determine how they vary. Then changes between them are subtracted from one another and corrected for noise. That should leave (more or less) the change that occurred between the images.

There are countless ways to observe changes between images. In medicine, one can monitor the growth of malignancies by evaluating images of a tumor, at different times, ranging from weeks to months to years. But there are also less-obvious applications for measuring changes, like when the amount that something changed is a matter of political or international importance.

For example, the progress of forest deforestation, which is measured by looking at the changes between two images of a forest, could have vast impacts on cap-and-trade policies, agreements in which billions of dollars are at stake. Therefore, in order to make sensible decisions based on changes in images, scientists need to know what they're measuring and why.

Methods for Good Change Detection

It's impossible to model an entire forest or an ocean, so imaging specialists choose a set of parameters, or dimensions, that they can use to characterize an image. For example, in a forest these dimensions might be the height or density of the trees.

The essence of detecting change is being able to recognize what doesn't change. Most changes are subtle, and most of the image doesn't actually change. Furthermore, it involves accounting for the aberrations in the image (clouds, snow, etc., while bearing in mind that the "noise" may contain significant data.

After the two images are generated, a process known as optical flow can be used to determine the relationship between the two images and therefore to create statistical models for changes in similar images. Optical flow is the process of asking what translation one can impose on each part of the image to create the next image.

The Pitfalls

However, there are a number of obstacles in measuring the meaningful changes, and these challenges vary depending on the type of image processing being used.

The amount of data one captures can create an image that is either too small to be meaningful, or so huge it's nonsensical. The challenge is to capture the right number of pixels (within instrumental and financial constraints, of course) so that the image is neither overwhelmingly hyperspectral nor underwhelmingly uniform. The change that an instrument detects may not signal an important change on the ground, after all. For example, the density of a radar signal may vary based upon the time of day, the time of year, and other factors.

When evaluating noise, a common pitfall is throwing out data points that fall within the range of what is considered extraneous information, or "noise." If there are multiple points in the noise range, after all, presumably those points would signify a meaningful cluster.

Then there comes the problem of whether the researcher should study the raw data versus the images that are mapped from the raw data. The latter option compounds the likelihood of error in detecting a change because there may have already been errors in the mapping of the image.

Finally, models, or the ways that data are processed into images, are not always perfect. Computational issues, poorly measured interference, imperfect algorithms, and the nonlinear nature of certain problems can all make it hard to generate an accurate image from the data collected.

There's no way to know that the model results are close to reality. And because the models are imperfect, there can be issues in characterizing the uncertainty of the answers.

Working Across Applications

To complicate matters further, each of these obstacles and their potential resolutions vary among scientific disciplines. One climatologist may use a program called MATLAB to convert data into an image, for example, while another will use a program called Fortran.

Furthermore, the type of data gathered varies by field. For example, environmental science may operate on a larger scale (like a forest), while biomedical sciences may operate on a smaller scale (like a tumor or a heart or a whole body). Furthermore, in some sciences, the data type is more or less ephemeral than in others—like heat waves versus rocks. For example, a climatologist would be more interested in measuring heat waves, while a geologist might be more interested in the composition of rocks. This discussion prompted the necessity to for some sort of basic, cross-disciplinary formula to describe the modeling of parameters as images. Our group chose to represent this basic formula in this way:

$$D = f(x,h) + n$$

Where D is the data (or image) constructed, f is the transfer function (or model), x is the parameter (such as height or biomass), h represents the hidden variables or nuisance parameters (such as atmospheric effects), and n is noise, or the parts of the image the researcher is not interested in measuring.

Depending on the task, however, some of these variables might be hard to define. In the task of supernova detection, for example, the parts of the image that aren't the supernova are things like other stars and cosmic rays. Therefore, the "h" is known and measurable. In land use, on the other hand, the final image comprises a variety of potentially confounding factors, such as clouds and shadows, none of which the researcher can predict. Therefore, the "h" is unknown.

Because of these differences in the variance of h, the model (f) for each of these disciplines can also vary. In astronomy, therefore, image detection tends to have a well-defined "f," while land use may have a poorly defined "f." These variances in "f" can make it challenging for scientists to work in one specific program or algorithm to detect changes between images. However, they may still benefit from studying the approaches taken in other disciplines, because the solutions may be applicable even if the data types are not.

Two Potential Solutions

There currently exists no canon of imaging science that can serve as a reference for image analysts across disciplines. Many scientists who actually perform image analysis were never academically trained in the practice, and instead learned on the job from others in their own profession.

In order to overcome these myriad obstacles, the IDR team proposes the creation of a common framework to detect changes in images across disciplines. Using methods that were developed in other fields would allow individual researchers to enhance their ability to detect meaningful change where they may have overlooked it previously.

The group proposed creating a textbook or online repository that would combine examples and frameworks for detecting important changes in images across all disciplines. It would also include software examples and tutorial datasets for the user to experiment with. This guide would serve as somewhat of an interdisciplinary "best practices" outline for image analysts so that they would not be circumscribed by the methods of their own disciplines. In this way, a geologist could see if a program or algorithm from medicine, for example, might suit one of his particularly challenging imaging tasks.

In 2009, the DVD rental service Netflix held a competition for whoever could find the best algorithm for predicting which movies users would like. In a similar vein, the IDR team proposes a multidisciplinary "image-change detection challenge." The challenge would provide a comprehensive dataset and a time series of images to analysts from any discipline. The analyst could then identify the features and estimate the dimensions of the change with his or her own tools or methods. Taking a cue from the "Netflix challenge" and other database contests, a cash prize could be awarded to those who successfully complete the challenge. By seeing the various approaches to the challenge, the original creators of the data set and images could see if there was a new or better approach to the change detection than the one they had been using

With these two solutions—the data challenge and online repository—image analysts would be better able to apply existing solutions to their current problem. That way, scientists from multiple fields would be provided with not only their own tools for detecting changes, but also those of their colleagues from other disciplines.

IDR Team Summary 4

Develop a telescope or starshade that would allow planetary systems around neighboring stars to be imaged.

CHALLENGE SUMMARY

The world in which we live is the only planet we know that harbors life. Is our planet unique? We have not yet found life on Mars, despite ample evidence of the existence of water, nor have we found evidence of life anywhere else in our solar system. A tantalizing possibility is that life may yet exist under the ice of the moons of Jupiter—yet there is no proof. Is there life elsewhere in the universe? If we were able to image planetary systems around neighboring stars, and in addition, characterize the surfaces and atmospheres of constituent planets, we would be one step closer to answering this question.

To date, more than 400 planets have been detected around other stars through a combination of radial-velocity techniques, transit experiments, and microlensing. Low-resolution spectra of a number of planets have also been found using the Hubble Space Telescope, the Spitzer Space Telescope, and a few ground-based observatories; in these cases, the planets have been objects unlike anything in our solar system, being mostly Jupiter-like planets in Mercury-like orbits. Images of several planetary systems have also been collected from the ground and space; these have shown planets in orbits much wider than even the bounds of our solar system and with planetary companions of extreme size, 3–20 times Jupiter's mass.

Planetary systems like our own around other stars are too small to be imaged by conventional telescopes. If we wanted to search around the nearest 150 stars, we would need a telescope with an angular resolution better than ~20 mas; this would allow us to distinguish objects such as Earth and Venus in solar system analogues at a distance of 15 pc from Earth. Our turbulent atmosphere limits ground-based telescopes to resolutions no better than 50 mas—even with the best available adaptive optics. Furthermore, the Hubble Space Telescope, with its 2.4 m mirror also has a resolving power no better than 50 mas. New advanced space telescopes are needed to image planetary systems similar to our own.

Beyond angular resolution limitations, a more difficult challenge is that planets are extremely faint as compared to the stars around which they orbit. An Earth-like planet would be about 10 billion times fainter than a Sun-like star when viewed at optical wavelengths, albeit somewhat brighter at infrared wavelengths—then only a factor of 10 million fainter. Because of this, scattered starlight within a telescope, caused by what would otherwise be negligible imperfections in mirror surfaces, can completely overwhelm the light from a planet. Telescopes must be significantly oversized compared to the required diffraction limited resolution so that planets could be seen beyond the glare of scattered starlight. Space telescopes with diameters of 8 m or more are needed to look for terrestrial planets around just the nearest dozen or so stars.

Building an 8-m optical space telescope is a formidable technical and engineering challenge. The largest telescopes on Earth are only slightly larger; namely the twin 10-m telescopes of the W. M. Keck Observatory. The largest telescope that can fit easily inside a launch vehicle is much smaller: only about 3.5 m in diameter. Innovative approaches to telescope design and packaging are therefore needed. In addition the telescope must have optics capable of suppressing starlight by a factor of 10 million to 10 billion—which is yet beyond the state of the art. Although this approach is certainly feasible with sufficient investment, it would provide images of only a handful of nearby planetary systems. Other innovative approaches have also been under study.

A potentially simpler approach might be to use a starshade to block starlight even before it enters the telescope, and have it an appropriate size and distance so that planet light could yet be seen. A starshade would need to be several 10's of meters in diameter and situated at several 10,000 km away from the telescope. This approach may greatly relax the engineering requirements on the telescope itself, but at the same time introduces other logistical challenges. It also would not significantly increase the number of planetary systems that could be imaged.

The limitations in angular resolution of a single telescope can be overcome if multiple telescopes are used simultaneously as an interferometer in a synthesis array. This provides an increase in resolution proportional to the telescope-telescope separation, not simply the telescope diameter. Since the late 1950s, radio astronomers have used arrays of radio telescopes for synthesis imaging, realizing that it would never be possible to build steerable telescopes larger than about 100 m (such as the National Radio Astronomy Observatory's Green Bank Telescope in West Virginia), nor fixed telescopes larger than ~300 m (the extreme example being Cornell's Arecibo telescope in Puerto Rico). Combining signals from separated telescopes is relatively straightforward at radio and millimeter wavelengths, because radio receivers with adequate phase stability and phase references are readily available. At optical and infrared wavelengths the problem is significantly more difficult, because of the increased stability requirements at these shorter wavelengths. Nonetheless, this approach seems to be a promising long-term path to imaging other planetary systems and finding life on other worlds.

An optical or infrared telescope array in space is also a formidable technical and engineering problem. Nonetheless, the required starlight suppression of a factor of 10 million (in the infrared) has been demonstrated in the lab. Telescope separations of up to 400 m are needed to survey the nearest 150 or so stars. The largest ground-based arrays, such as Georgia State University's Center for High Angular Resolution (CHARA) Array on Mount Wilson, California, have telescope separations of up to 300 m. However, atmospheric turbulence limits their sensitivity to objects brighter than 10–14th magnitudes. A space telescope array, above the atmosphere, would have a sensitivity limited primarily by the collecting area of each telescope, but there would be no single platform large enough on which to mount it. The telescopes would need to be operated cooperatively as a formation-flying array: this was for many years the baseline design of NASA's Terrestrial Planet Finder (TPF) mission. Although experiments in space have demonstrated rendezvous and docking of separate spacecraft, no synthesis array has yet been flown. There is no precedent for a mission like TPF.

Key Questions

• What innovative new ways and approaches might there be from other disciplines that could reduce the cost and increase the science of a planet-imaging mission?

- How might NASA's Human Spaceflight Program be used to build new observatories in space?
- How should NASA best invest in technology to enable future planet-imaging missions?

Reading

Fridlund M, et al. The astrobiology habitability primer. *Astrobiology* 2010;10:1-4. Abstract accessed online June 15, 2010.

Hand E. Telescope arrays give fine view of stars. *Nature* 2010;464:820-1. Accessed online June 15, 2010.

Oppenheimer BR and Hinkley S. High-contrast observations in optical and infrared astronomy. *Ann Rev Astron Astrophys* 2009;47:253-89. Abstract accessed online June 15, 2010.

Schneider J. Extrasolar Planets Encyclopaedia: Interactive Extra-solar Planets Catalog, with all planets detected by imaging, with references. Accessed online June 15, 2010.

IDR TEAM MEMBERS

- Supriya Chakrabarti, Boston University
- Richard A. Frazin, University of Michigan
- Jennifer D. T. Kruschwitz, JK Consulting
- Tod R. Lauer, National Optical Astronomy Observatory
- Peter R. Lawson, California Institute of Technology
- Timothy P. McClanahan, NASA Goddard Space Flight Center
- Richard G. Paxman, General Dynamics-Advanced Information Systems
- Lisa A. Poyneer, Lawrence Livermore National Laboratory
- George R. Hale, Texas A&M

IDR TEAM SUMMARY

George R. Hale, NAKFI Science Writing Scholar, Texas A&M

For years people have looked at the sky and wondered if there were other Earths out there. It wasn't until about 15 years ago that we knew that other stars in the cosmos had planetary companions. Today we know about the presence of hundreds of extrasolar planets, but we've only actually seen a handful of them.

The vast majority of extrasolar planets have been discovered using indirect methods such as the radial velocity method or seeing transits of the planet in front of its star. The reason that only a few planets have been directly imaged is because they are incredibly difficult to see. One well-worn analogy is that finding a planet orbiting a distant star is like looking for a firefly next to a searchlight in New York while you're in San Francisco.

This analogy highlights two of the major difficulties faced when imaging such planets. First, they are very far away and therefore hard to resolve without a very powerful telescope, and second, a star's glare "drowns out" the planet's light. Blocking out that glare can be done with a coronagraph or with a more complicated device called a starshade.

Can We Get a Different Question?

Originally, IDR team 4's challenge was to develop a telescope or starshade that would allow planetary systems around neighboring stars to be imaged. Significant work in this areas is either already under way or planned for the near future. Hardware has been considered but isn't in line with the imaging science–related theme of the conference. As a result, team 4 decided to reframe the challenge of imaging extrasolar planets in a more topical way in the hopes that they might make more progress instead of rehashing existing information. The new question is: "How do we apply imaging science to detect and characterize exoplanets?"

Why Are We Interested?

Over the next few years, several new instruments like Kepler, SPHERE, the Gemini Planet Imager (GPI), and PICTURE will come online and begin to collect vast amounts of data. This coming flood of information means we need better imaging methods now, not in 5 to 10 years.

That the hundreds of planets we have found thus far have been nothing like what we expected to find is another motivation for improving imaging. Because of limitations with the radial velocity method, the planets we have found are huge (larger than Jupiter) and have close orbits (closer than Mercury is to the sun). The ones we really want to find—those that my be habitable—are going to be smaller and farther away.

Problems to Overcome

Direct imaging doesn't have the same limitations that the radial velocity method faces, which means it should be possible to find the planets astronomers want to find. But it does have its own challenges such as brightness differences, atmospheric turbulence, and image noise.

First and foremost, stars and the planets that orbit them have vastly different brightness. For example, the planets in HR 8799—one of the handful of directly imaged systems—are several thousand times dimmer than their parent star. Finding these planets was hard, but imaging a smaller and farther out planet would be even harder. Seeing a planet the size of Earth would require a reduction by a factor of 10^9. In other words, it would be a billion times dimmer than the star it orbits.

Atmospheric turbulence has been and continues to be a headache for astronomers worldwide. A space telescope is one way to get around turbulence, but such instruments are very expensive and pose their own engineering challenges. Another method is through the use of adaptive optics (AO). AO systems use deformable mirrors and computer software to attempt to compensate for turbulence, but they aren't perfect. Even the best AO systems leave some noise in the image, which takes the form of streaks or "sparkles" in the background. This noise makes finding something small and faint like a planet exceptionally difficult, so it is desirable to remove or cancel out the interference.

Detection and Categorization

What's needed is a way to find out what's limiting imaging performance. To begin understanding how to do this we have to consider the data. Image data from telescopes comes in the form of data cubes. These cubes have spatial data (x and y coordinates), and data about light wavelength and polarization, and a series of cubes taken at different time intervals can give researchers temporal data. Wavelength and polarization data are of particular interest because how light is polarized can tell about the presence and composition of dust around a star, and the light's spectrum can give clues about the presence of substances like water, methane, and chlorophyll—signs of life.

But before we can look for signs of life we have to first find the planet. Understanding how the noise in the image behaves over time, when the image moves, and how different spectra of light behave is important. To understand what the sparkly background looks like, picture an elongated, ice cream cone–shaped streak that smears from red to blue. A planet in this image would be a single point and would only show one spectrum. As for motion, when a telescope fixes its gaze on a star, the noise stays in place while planets move.

Inquiring Minds Want to Know

How much better can we do? That's the kernel of IDR team 4's question. What are the different modeling approaches, and what is the best statistical framework? How do we choose which is best? A method that takes two hours to process one data cube is of little benefit. Image processing needs to be done in real time, and improvements have to be weighed against their costs in time and money.

How much better should scientists try to do is a related question. Improvements in exoplanet images can be thought of as running on a continuum, with doing nothing to the image at all on one end and having an ideal linear observer on the other. Currently, researchers are using methods like angular differential imaging (ADI), spectral differential imaging (SDI), and locally combined combination of images (LOCI) to eliminate noise. These techniques are an improvement but are only ad hoc methods.

What we must know is the source of fundamental error. Better understanding the source of error is the first step in developing an algorithm to correct for it. Such an algorithm would also need to be adaptive, changing when an AO system does something unexpected. One source of such information could come from AO systems themselves. AO systems produce corrected images, but they also produce a constant stream about the atmosphere and systematic data, that is, what the AO system is doing. The question is: Can we exploit this? This auxiliary data could hold the key to developing a better image processing algorithm.

Lastly, it's worth considering how improved imaging could inform hardware decisions for future instruments. For instance, GPI will go online soon. Had better image correction algorithms been developed five years ago, would the instrument look any different? Will improved algorithms lead to better images and better hardware?

IDR Team Summary 5

How can we extend the domain of adaptive optics and adaptive imaging to new application, and how can we objectively compare adaptive and non-adaptive approaches to specific imaging problems?

CHALLENGE SUMMARY

Adaptive optics has revolutionized ground-based optical astronomy, and it has found important applications in ophthalmology, medical ultrasound, optical communications, and other fields where it is necessary to correct for a phase-distorting medium in the propagation path. Most often, an adaptive optics system uses an auxiliary device such as a Shack-Hartmann wavefront sensor to characterize the instantaneous distortion, and it then uses a deformable mirror or other spatial phase modulator to correct the distortion in real time.

Adaptive imaging is a broader term than adaptive optics. It refers to any autonomous modification of the imaging system to improve its performance, not just correcting for phase distortions and not necessarily relying on auxiliary non-imaging devices such as wavefront sensors. One paradigm is to collect a preliminary image, perhaps with a short exposure time and relatively limited spatial resolution, then to use this information to determine the best hardware configuration and/or data-acquisition protocol for collecting a final image. Alternatively, the process can be repeated iteratively, and the optimum system for any current acquisition can be based on all previous acquisitions. In either case, the imaging information used to control the adaptation can be from the system being adapted or from some other imaging system, and it can even be from an entirely different imaging modality.

An example of the latter paradigm derives from a popular multimodality approach to medical imaging in which a functional imaging modality such as PET (positron imaging tomography) is combined with an anatomical modality such as CT (computed tomography); normally the images are either superimposed or read together by the radiologist, but it is also possible to use the information from one of the modalities to control the data acquisition in the other modality (Clarkson et al., 2008; referenced under Reading).

The goal of either adaptive optics or the more general adaptive imaging is to improve the quality of the resulting images. Most often the quality has been assessed either in terms of image sharpness or subjective visual impressions, but it is also possible to define image quality rigorously in terms of the scientific of medical information desired from the images, which is often referred to as the task of the imaging system. Typical tasks in medicine include detecting a tumor and estimating its change in size as a result of therapy. In astronomy, the task might be to distinguish a single star from a double star or to detect an exoplanet around a star. The quality of an imaging system, acquisition procedure, or image-processing method is then defined in terms of the performance of some observer on the chosen task, averaged over the images of many different subjects.

The methodology of task-based assessment of image quality is well established in conventional, non-adaptive imaging (although some computational and modeling aspects will be explored under IDR Team Challenge 2), but very little has been done to date on applying the methodology to adaptive systems. Barrett et al. (2006) discusses task-based assessment in adaptive optics, and Barrett et al. (2008) treats the difficult question of how one even defines image quality, normally a statistical average over many subjects, in such a way that it can be optimized for a single subject. Much more research is needed on image quality assessment for all forms of adaptive imaging.

Key Questions

- What imaging problems are most in need of autonomous adaptation? What information, either from the images or from auxiliary sensors, is most likely to be useful for guiding the adaptation in each problem?
- For each of the problems considered under the first question, what are the possible modes of adaptation? That is, what system parameters can be altered in response to initial or ongoing image information?
- Again for each of the problems, how much time is available to analyze the data and implement the adaptation? What new algorithms and computational hardware might be needed?

- What new theoretical insights or mathematical or statistical models are needed to extend the methodology of task-based assessment of image quality to adaptive imaging?

Reading

Barrett HH and Myers KJ. Foundations of Image Science. John Wiley and Sons; Hoboken, NJ, 2004.

Barrett HH, Furenlid LR, Freed M, Hesterman JY, Kupinski MA, Clarkson E, and Whitaker MK. Adaptive SPECT. *IEEE Trans Med Imag* 2008;27:775-88. Abstract accessed online June 15, 2010.

Barrett HH, Myers KJ, Devaney N, and Dainty JC. Objective assessment of image quality: IV. Application to adaptive optics. *J Opt Soc Am A* 2006;23:3080-105. Abstract accessed online June 15, 2010.

Clarkson E, Kupinski MA, Barrett HH, and Furenlid L. A task-based approach to adaptive and multimodality imaging. *Proc IEEE* 2008;96:500-511. Abstract accessed online June 15, 2010.

Roddier F., Ed., Adaptive Optics in Astronomy. University Press; Cambridge, U.K., 1999.

Tyson RK. Introduction to Adaptive Optics. SPIE Press; Bellingham, WA, 2000.

IDR TEAM MEMBERS

- Thomas Bifano, Boston University
- Liliana Borcea, Rice University
- Miriam Cohen, University of California, San Diego
- Jason W. Fleischer, Princeton University
- Craig S. Levin, Stanford University
- Teri W. Odom, Northwestern University
- Rafael Piestun, University of Colorado at Boulder
- Hongkai Zhao, University of California, Irvine
- Lori Pindar, University of Georgia

IDR TEAM SUMMARY

Lori Pindar, NAKFI Science Writing Scholar, University of Georgia

From the Heavens to Earth: Adaptive Optics and Adaptive Imaging

From the time Galileo first looked through his telescope to catch a glimpse at the night sky, the quest for how to best see past the visual limitation of the human eye gained a fervor that has yet to lose momentum. In recent history, adaptive optics helped revolutionize the field of astronomy as telescopes became more powerful and better able to image through Earth's atmosphere to see the celestial bodies in the universe. The benefits of adaptive optics did not stop there and have found practical application in fields such as medical imaging like ultrasound and visualization of the retina. Adaptive optics fits within the realm of adaptive imaging, a general term to describe techniques in which the ultimate goal is to improve the quality of the pictures received by those who use these resulting images for research, analysis, and diagnosis.

Therein lies the beauty and challenge of imaging and optics when we seek to make them adaptive. It introduces a set of issues ranging from technology and algorithms to how quality of an image best suits the problem proposed across multiple fields of study. However, imaging science is built on progression, and system performance is key in advancing the field, accruing accurate data, and ensuring that the best methods are being employed to benefit society as a whole.

A Picture Is Worth A Thousand . . . Pre-Detection Corrections

As IDR team 5 began its task, the first discussion revolved around creating a unified interdisciplinary understanding of what adaptive optics and adaptive imaging mean and what mechanisms and technologies use various correction technologies to improve imaging. The final consensus was that adaptive optics specifically deals with the correction of a wave front in an optical system with feedback and iteration. For this type of imaging system, there are multiple sources of aberration that impede picture quality such as reduction of contrast, resolution, or brightness. However, by adjusting for these sources of aberration (or adapting the system), the distortions can be counteracted.

Adaptive imaging, while including adaptive optics, also extends beyond to the realm of optimization. Adaptive imaging, for this challenge's purpose, can be thought of as improvements to an image, while adaptive optics is concerned with the actions taken to compensate for aberrations that affect the image in the initial capture.

New Applications

How can the domain of adaptive optics and adaptive imaging be extended to new applications? The IDR team immediately saw that the prob-

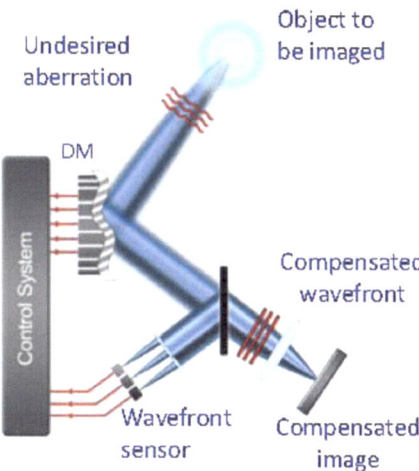

FIGURE 1. The Classic Application of Adaptive Optics: Correction of wavefronts in an optical system in which medium aberrations are compensated through feedback to a deformable mirror (DM). Image courtesy of T. Bifano.

lem does not lie in how the field could be extended but in issues within the limitations of adaptive optics and imaging itself, as well as improvements to various fields that can be gained by use of adaptive optics. Questions regarding increased data correction, measure, and speed were introduced in order to better assess those applications for which adaptive optics and adaptive imaging are most amenable. For example, in biological tissue samples, when light or radiation is passed through a sample, there needs to be a proper balance so that the tissue can be seen without destroying the integrity of the sample. Thus, there are certain limitations that need to be taken into account depending on the field as well as what the person is looking for; therefore optimization and gaining the sharpest image is not always the simple answer when what you are looking for could be obscured, ironically, by making an image better.

To practically assess the problem, the team focused primarily on adaptive optics but did give some attention to adaptive imaging. Adaptive optics addresses the issues of aberrations, often unavoidable, such as ground-based telescopes that compensate for Earth's atmosphere in order to see into outer space, or retinal imagers that examine the cornea of the eye. However,

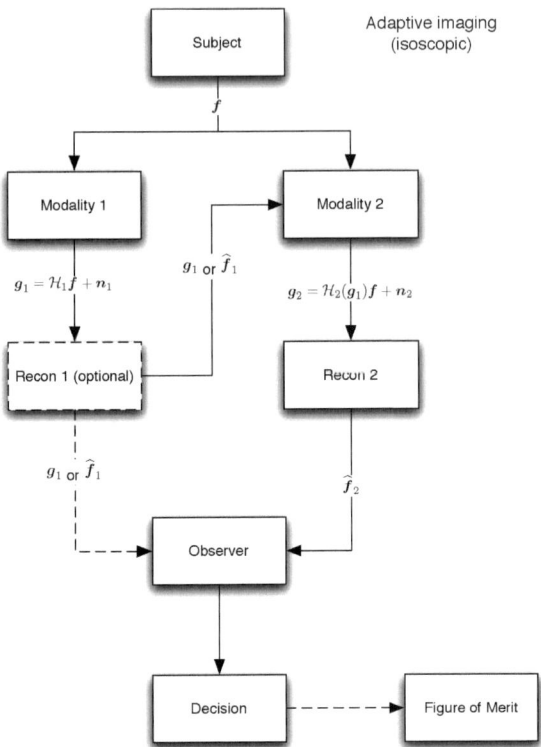

FIGURE 2. Adaptive imaging is an active means of improving performance in an imaging system that includes adaptive optics and task-based control such as auto-focus. Image courtesy of H. H. Barrett.

because these media and the aberrations they produce are known, the instruments measuring them can be fixed to continually adjust and produce an accurate reading or measure of the object of study. The usual objective with adaptive optics is to reach a diffractive limit within a linear domain. So within an optical imaging system like a microscope or cameras, its optical power is only as strong as its imperfection in the lenses or alignments. For example, a 10 megapixel digital camera with autofocus will only provide images of 10 megapixel quality—which is appropriate if that is all you need but not so if you strive for 12 megapixel images. Although the scale of adaptive optics is much larger than of an ordinary digital camera, the example illuminates the limitations that technology reaches when dealing with tools that continuously correct for imperfections within a system. In order to reach this diffraction limit, adaptive optics attempts to correct for such imperfections so that these systems can perform at their best, but it remains a task that is still difficult to accomplish—unless one is dealing with a space-based telescope that does not have to contend with atmospheric aberrations. However, the team took a novel approach in extending adaptive optics to the nonlinear domain and the methods in which to improve resolution in order to create a step-change outcome in applying adaptive optics to new arenas.

Emerging Applications in Adaptive Optics and Adaptive Imaging

1. Fixing (intentional) aberrations
   a. Physical or algorithmic
   b. Cubic phase plates; deliberately put in and then post process to remove so system becomes insensitive to defocus
2. Adapting nonlinear systems
3. Volume imaging: In the instance that phase aberrations become amplitude aberrations, use adaptive optics as a diagnostic tool
   a. Thick tissue (limits depth)
   b. High scattering media
   c. Patients
   d. Neuroimaging
   e. Lungs (use to compensate for motion)
4. Employing multiscale adaptive optics
   a. Molecular resolution at arbitrary depths in scattering media (current 0.5 mm)
5. Getting better (static, background) noise properties
   a. Put a deformable mirror in two-photon microscope

The team also addressed the question, "If more was known about the media, how would that change the data acquisition process?" This led to the following ideas:

6. Applying adaptive optics to plasmonics to mold dynamically (nanoscale; adaptive imaging)
7. Using beam forming as an analog to adaptive optics
   a. Focus a bright beam at an object in an effort to make it sharper

8. Improving acqusition through optics and other energy sources (acoustic, heat, etc.)
9. Angular dependent measurements, scattering kernel
10. Preliminary probing to get sparse information and then adapting accordingly for higher resolution
11. Foveated imaging. High resolution at point(s) of interest in a wide field image (by sculpting wave front to have tilt with spatial modulator, change focal length of optical elements with translating optics, deformable mirrors)
    a. Practical for use in digital cameras, unmanned autonomous vehicles.
12. Compressive adaptive imaging
    a. Imaging with past measurements in order to increase data speed and create higher resolution images

In extending adaptive optics and adaptive imaging to new systems, the prior conceptual list can be assessed by various fields and possibly build new relationships that are interdisciplinary and pose mutual benefit to implementing and using improved adaptive imaging systems.

To Adapt or Not Adapt . . . Is That the Question?

The second part of the task dealt with how to objectively compare adaptive and non-adaptive approaches to specific imaging problems. In order to tackle this problem the group asked how one can compare solutions that both adaptive and non-adaptive imaging create with the idea that if focusing a microscope is adaptive optics (on a certain level), despite there not being a multi-step system of inputs improving the image. Therefore, the second question spurned further questions.

Comparing Adaptive and Non-Adaptive Approaches: Further Questions

1. Are these fundamentally different imaging systems?
2. Do these systems actually require different metrics?
3. Does the specific imaging problem depend on the task?
4. How would you choose to manipulate data?
5. Issue of task-based metric and adaptive systems.
   a. Receiver observer based on ensemble of objects but the system becomes completely different under a different operator.

6. Are there two different ways of adapting to compare the result?
7. Inverse problem; make problem convex—optimize an intermediate process but only when prior information is present.
8. Issue of creating a phantom that is appropriate to task (e.g., resolution, noise, or signal)
9. Correlation between feedback system and task; control metric and imaging metric—are they different? How does control loop relate to task? Can a surrogate figure of merit (FOM) be calculated on basis of individual images? If so, it can be correlated with task performance, and can it use on-the-fly adaptive imaging?

Key Challenges of Domain Extension with Adaptive Optics and Adaptive Imaging

Taking into consideration the ideas for new applications and the questions that arose from the discussion, the team decided to address two key challenges that pose a foreseeable impediment to implementation of adaptive systems. The first is the problem of using adaptive optics to image in and/or through a 3-D volume. Turbulence is one aspect to contend with, but add a long distance or wider field of view and the problem becomes a bit more layered and difficult to assess. Also, imaging through thicker samples, in biological imaging for example, will induce aberrations that are difficult to contend with because the amount of scatter will be increased. Therefore an inverse problem is introduced in scenarios where feedback may not be readily available to adapt (correct) image capture. For adaptive imaging, the problem involves imaging in and through highly scattered media. This problem is especially relevant in medical imaging through turbid media such as bodily tissue.

This further brings about computational issues such as sensor speeds and how algorithms based on linear models should be adjusted to image in volumes. Also, in a task-based sense, there needs to be an adaptive step to improve figures of merit so that computations are accurate and involve all aspects of alterations and aberrations within a system.

Future Steps: A Glimpse into the Periphery

In order to create a method of implementation to extend adaptive optics and adaptive imaging to new spheres, the team saw the need to begin by analyzing several key components of the adaptive system so that mutual benefit in interdisciplinary fields could be gained for future imaging solutions.

These ideas include solving the inverse problem as well as creating an updated model for adaptive imaging to correct for aberrations or changes within a system. Also, there is benefit in extracting information from 3-D measurements to better know what exists (and thus, what affects the measurement) between the point of interest and the imaging technology. Media also needs to be approached differently depending on how turbulent or turbid they are, and physical system corrections need to be implemented. Such implementations are in the future because they are application-dependent, and how to actually correct the physical system is a question in and of itself.

Although improving the quality of images and their systems is an important goal, future tools, systems, algorithms, and other components of adaptive imaging and adaptive optics are yet to be designed and implemented. However, the team envisioned a future that can be brought from the peripheral cusp of ingenuity and into focus for future generations of imaging science applications.

IDR Team Summary 6
What are the tools and validation methods required to develop clinically useful non-invasive imaging biomarkers of psychiatric disease?

CHALLEGE SUMMARY

"Biomarker" is a term often used in the biomedical disciplines for a characteristic that can be used as an indicator of some biological condition or outcome that is ultimately of interest but difficult to ascertain directly, at least with respect to applications in human disease. A number of implicit and explicit definitions of "biomarker" are in common circulation. There are also quite different uses of the term in other disciplines (National Research Council report, *Opportunities in Neuroscience for Future Army Applications*, referenced in "Suggested Reading"). To avoid confusion, the IDR Team may wish to adopt the following definition, published by the Biomarkers Definitions Working Group of the National Institutes of Health (Atkinson et al., 2001, p. 91, referenced under Reading):

> Biological marker (biomarker): A characteristic that is objectively measured and evaluated as an indicator of normal biological processes, pathogenic processes, or pharmacologic responses to a therapeutic intervention.

Therefore, biomarkers should be proven surrogate endpoints that can accurately predict clinical endpoints such as how a patient feels, functions, or whether or not the patient survives. In the field of psychiatry, diseases are categorized and identified based on clinical symptoms that may not specifically reflect the underlying pathophysiology present. Biomarkers have the potential to probe physiological processes to provide quantitative methods for differentiating illnesses with similar clinical symptoms resulting from unique neurobiological mechanisms, such as schizophrenia and other forms of psychosis. Furthermore, many psychiatric diseases have early asymptomatic or transitional phases that can last years. In the case of schizophrenia, many patients experience several pre-psychotic phases prior to full onset of symptoms. Novel biomarkers could potentially identify those individuals most likely to progress to the illness in order to target preventive or early treatment. In addition to prediction and early detection, longitudinal biomarker measures may be used to assess the trajectory of the disease process. Those biomarkers that are sensitive to changes in the trajectory caused by certain drugs could be used to identify potential responders to specific pharmacological interventions (personalized medicine).

Currently there are many potential biomarkers in development for diseases such as Alzheimer's disease, depression and schizophrenia. In depression, measures of quantitative electroencephalogram (EEG) concordance are being investigated. Potential biochemical markers include measurement of amyloid plaques and other proteins or hormones in cerebrospinal fluid or plasma. A large body of research is also focused on the search for genetic markers of these diseases. Currently, these markers have been promising in identifying individuals who are at high risk for illness; however, many lack specificity to predict those who will progress to full symptoms among the vulnerable individuals.

Neuroimaging provides several non-invasive tools for the determination of potential psychiatric biomarkers. Structural imaging such as anatomic magnetic resonance imaging (MRI) can detect lesions and atrophy, as well as cerebrovascular disease, associated with a variety of psychiatric illnesses. Functional imaging such as positron emission tomography (PET) has been used to find distinctive patterns of glucose hypometabolism in dementia. Also, changes in brain function in response to specific cognitive tasks measured with functional MRI (fMRI) may be indicators of disease. White matter tract integrity can be measured with MRI diffusion tensor imaging and has been found to correlate with drug response in patients with depression and in early stages of schizophrenia. Another MRI technique, MR spectroscopy is useful in measuring biochemical levels in a single voxel or across the brain to assess integrity of specific metabolic pathways that may be altered in disease. These and other imaging biomarkers may be combined with complementary physiological and biochemical measures for the most sensitive and specific indicators of illness.

The safe and non-invasive nature of many of these imaging techniques such as MRI make them attractive for early and repeated screenings of large populations. However, while these neuroimaging biomarkers of psychiatric disease are promising, there are many obstacles that must be overcome in order to make them clinically feasible. These include, but are not limited to, the following: imaging and processing methods must be standardized and repeatable; results must be validated and interpretable in relation to standard norms of brain maturation and aging; drug effects on potential measures must be fully elucidated; and studies must be performed with large populations of well-characterized patients and at-risk individuals. The challenge here is to identify all of the qualifications of clinically useful imaging biomarkers for psychiatric disease and the tools and methods required to develop these efficiently.

Key Questions

- How can imaging biomarkers be used to demonstrate neurobiological mechanisms of disease?
- How can they be used to determine disruptions in connectivity across brain networks that may be the underlying cause of psychiatric brain disorders?
- What new technologies are required to allow validation and increase predictive power of imaging biomarkers of psychiatric diseases?
- What methods/qualifications should be used in determining which imaging modality is most useful for the given application? How do we jointly optimize the biomarker and the imaging system?
- How can imaging biomarkers be most efficiently utilized in conjunction with more invasive biomarkers?
- What can we learn from on-going large mutli-site studies such as the ADNI study (http://www.adni-info.org/Home.aspx)?

Reading

Atkinson AJ Jr., Colburn WA, DeGruttola VG, DeMets DL, Downing GJ, Hoth DF, Oates JA, Peck CC, Schooley RT, Spilker BA, Woodcock J, and Zeger SL. Biomarkers and surrogate endpoints: Preferred definitions and conceptual framework. *Clin Pharm Ther* 2001;69(3):89-95. Abstract accessed online June 15, 2010.

Cedazo-Minguez A and Winblad B. Biomarkers for Alzheimer's disease and other forms of dementia: clinical needs, limitations and future aspects. *Exp Gerontol* 2010;45:5-14. doi: 10.1016/j.exger.2009.09.008. Accessed online June 15, 2010.

Correll CU, Howser M, Auther AM, Cornblattt BA. Research in people with psychosis risk syndrome: a review of the current evidence and future directions. *J Child Psychol Psychiatry* 2010;51(4):390-431. doi: 10.1111/j.1469-7610.2010.02235.x. Abstract accessed online June 15, 2010.

Kumar A, Ajilore O. Magnetic resonance imaging and late-life depression: potential biomarkers in the era of personalized medicine. *Am J Psychiatry* 2008;165(2):166-8. doi:10.1176/appi.ajp.2007.07111771. Accessed online June 15, 2010.

National Research Council. Opportunities in neuroscience for future army applications. National Academies Press; Washington, DC, 2009.

Pantelis C, Yucel M, Bora E, Fornito A, Testa R, Bewer WJ, Velakoulis D, Wood SJ. Neurobiological markers of illness onset in psychosis and schizophrenia: the search for a moving target. *Neuropsychol Rev* 2009;19:385-98. doi: 10.1007/s11065-009-9114-1. Accessed online June 15, 2010.

IDR TEAM MEMBERS

- Owen T. Carmichael, University of California, Davis
- Dennis W. Choi (IOM), Simons Foundation
- Marcel Adam Just, Carnegie Mellon University
- Linda J. Larson-Prior, Washington University in St. Loius
- Carolyn C. Meltzer, Emory University
- Andrew B. Raij, University of Memphis
- A. Ravishankar Rao, IBM Research
- James M. Rehg, Georgia Institute of Technology
- Bruce R. Rosen, Massachusetts General Hospital
- Kathleen M. Raven, University of Georgia

IDR TEAM SUMMARY

Kathleen Raven, NAKFI Science Writing Scholar,
University of Georgia

Brain images of neurological and psychiatric disorders are needed to help research, make diagnoses, track disease progression, and monitor treatment. One problem that neurological researchers face is how to detect a disease as early as possible. The safe, non-invasive nature of imaging technology available today makes them attractive for repeated screenings of large populations in order to identify *images* that mark the onset or presence of neurological diseases. In order to create image biomarkers, researchers would collect data using technology such as magnetic resonance imaging (MRI) and positron emission tomography (PET). Currently scientists and physicians can detect clear neurobiological changes visible from a stroke, for example. In the future, similar changes in the brain as a result of depression may be visible.

Image biomarkers would complement already established biochemical and genetic biomarkers for diseases such as Alzheimer's and depression. For example, a biochemical marker of Alzheimer's disease is the presence of amyloid plaque in a patient's brain visible through MRI scans. The trouble with biochemical and genetic markers is that they work best only in high-risk individuals; many lack the specificity needed to predict who will progress to have full disease symptoms later.

Although image biomarkers of psychiatric diseases are promising, many obstacles must be overcome in order to make them clinically feasible. These include, but are not limited to, the following: the imaging and processing methods must be standardized and repeatable; the results must be validated and able to be interpreted in relation to standard norms of brain maturation and aging; the drug effects on potential measures must be fully elucidated; and studies must be performed with large populations of well characterized patients and at-risk individuals. The challenge here is to identify all the qualifications of clinically useful imaging biomarkers for psychiatric disease and the tools and methods required to develop these efficiently. An interdisciplinary research team (IDR) at the National Academies Keck *Futures Initiative* Conference on Imaging Science debated these and other challenges surrounding the tools and validation methods required to develop clinically useful non-invasive imaging biomarkers of neurodegenerative and psychiatric disease.

For the purposes of IDR team 6's discussion, the group agreed to the biomarker definition provided by the Biomarkers Definitions Working Group of the National Institutes of Health:

> A characteristic that is objectively measured and evaluated as an indicator of normal biologic processes, pathogenic processes, or pharmacologic responses to a therapeutic intervention. (Atkinson et al., 2001, p. 91)

The group conceded that image biomarkers have several distinct applications. These include identification of the disease state, tracking disease progression, and evaluation of therapeutic response. Because of the brain's structural, anatomical, and functional complexity, the group advised against searching for single image biomarkers. Instead, a biomarker panel should be used. The panel should incorporate data from structural imaging (e.g., MRI and MR spectroscopy) and functional imaging (e.g., PET scans). A successful biomarker panel would also take into account changes in the brain's anatomy, physiology, metabolism, electrophysiology, and neurochemistry over time. Examples of indirect biomarkers—these could also be thought of as "proto-biomarkers"—could be markers that predict the disease course or therapeutic response. The team cited blood pressure as a classic indirect biomarker—although unrelated to the field of psychiatric diseases—in the prevention of cardiovascular disease.

One of the main concerns shared by the group in determining image biomarkers is how to identify relevant information within brain structure as image resolution becomes higher and higher. The team predicted a massively expanding role for structural imaging as image resolution approaches the 10μ (micron) scale. Researchers will contend with even more image information and will need to make additional decisions about what is useful and useless. However, greater magnification of brain structure could potentially shrink the continuum between "neurobiological" and "psychiatric" diseases.

One team member who specializes in autism research suggested that data points gathered from behavioral patterns should be strongly correlated with image biomarkers. Autism research has long been tied to behavioral markers, such as the direction and length of a child's gaze on the moving lips of a person talking. Novel sociological means of capturing such information could include such devices as goggles for toddlers designed to track eye gaze to conversation-monitoring devices on mobile phones.

Imaging biomarker tools and methods should take into account temporal changes in the brain across time spans as short as image measurement time to as long as years. Longitudinal data can help researchers distinguish significant from insignificant changes in brain images. Similarly, the team suggested the assessment of "microstates" beyond just arousal, or non-resting/sleeping, state of the brain. The tracking of these states could lead to predictive or disease markers and possibly guide therapeutic treatment.

Additional recommendations for imaging biomarkers made in the final presentation are as follows:

• Explore the role for public-private partnerships (e.g., the Foundation for the National Institutes of Health and the European Union's Innovative Medicine Initiative) in driving the coordinated development of optical imaging technology and novel molecular probes
• Emphasize methods for imaging through skull and gaining portability
• Combine structural and functional modalities to develop increasingly precise, multidimensional baseline images of regions and pathways altered in psychiatric diseases (e.g., reward, mood, and theory of mind)

• Assess both highly selected and representative—or comorbid—populations in search for biomarkers
• Emphasize adaptation of large-scale informatics, computational, and feature extraction approaches to novel forms of imaging and behavioral datasets (e.g., graphs)

The group identified priorities within two types of biomarkers that could have significant implications for public health and would help, through early detection and treatment, reduce the cost of health care. Within predictive biomarker research, much work remains to be done on the placebo response to antidepressant therapy. Identifying biomarkers associated with the emergence of earliest symptoms in presymptomatic individuals would allow earlier intervention. Opportunities exist to determine the predictive markers of weight gain, which is the most common side effect associated with atypical antipsychotic agents. Therapeutic biomarker priorities include mood disorders, deficient social interactions, symptoms of schizophrenia, and alcoholism.

In conclusion, the group cautioned against confusing biomarkers of disease with *byproducts* of disease. A common example is the formation of amyloid plaque found in patients diagnosed with Alzheimer's disease. It is unclear if the plaque is an indicator of the disease or the result of a disease mechanism. A recurring theme throughout the discussion was the important role that computer software developers currently have, and will continue to have, in processing new information gathered from the ways discussed above. As more information is collected, the sophistication of computational methods and algorithms to make sense of data will, by necessity, increase.

IDR Team Summary 7

Find novel ways to use imaging methods to improve the treatment of diseases.

CHALLEGE SUMMARY

The development of treatments for human diseases has followed models and methods that have changed very little during the past 50 years. Capitalizing on the power of imaging technologies offers an opportunity to improve these models and methods and to make the search for improved treatments more efficient and the treatments themselves more efficacious. Traditionally the search for treatments goes through stages: target identification, compound synthesis and screening, evaluation in animal models, phase 1-4 testing, and assessment of outcomes such as efficacy or side effects. The challenge to imaging technology is to find ways with which this search could be improved.

Concepts That Might Be Useful to Address the Challenge

Improving Dose-finding Strategies with Positron Emission Tomography (PET)

Traditionally, dose-finding studies have relied on a relatively crude trial-and-error approach in which multiple doses are used, and symptom improvement has been the measurable target. The development of methods to image targets that are more primary to the disease process offers the possibility to improve dose-finding methods and outcome assessment. For example, labeled ligands have been developed that bind to neuroreceptors for neurotransmitters that are thought to be overactive or underactive in brain diseases (e.g., dopamine in schizophrenia, Parkinson's disease; serotonin in mood disorders) and therefore are conceptualized as more precise treatment targets. These ligands (e.g., [^{11}C]raclopride for D2 receptors, [^{18}F]setoperone for 5-HT$_2$ receptors) can be used to measure the occupancy of receptors induced by varying doses of medications, and the degree of occupancy can then be correlated with level of symptom improvement or side effects to determine the level of receptor occupancy required for optimal treatment. This approach is now widely used in order to determine the optimal doses of some psychoactive drugs. The number of available ligands that are U.S. Food and Drug Administration (FDA)-approved is limited, however, and so the development of new and better ligands for drugs of many types continues to be a significant challenge.

For many years [^{18}F]fluorodeoxyglucose-2-deoxy-D-glucose (FDG) has been used to identify the location and size of cancer lesions and to monitor their response to treatment using PET. FDG, while useful, is also relatively crude and nonspecific. The application of imaging technology to monitoring treatment targets can be substantially enhanced if investigators develop new amino acid ligands (tyrosine, methionine, thymidine) that aim at more specific targets, such as hormones (e.g., receptors for estrogen, testosterone) or substrates involved in protein or nucleic acid synthesis.

Novel Technologies to Improve Treatment by Manipulating Intracellular Activity

Most current applications of imaging technology to improving treatment examine activity on a large scale: systems, organs, lesions, etc. A new technology has recently emerged that permits imaging at the intracellular level and the ability to manipulate cellular function. This technology is referred to as "optogenetics" because it uses light-responsive proteins derived from algae (channelrhodopsin and halorhodopsin) that can be used to manipulate neuronal firing by opening or closing ion channels. Channelrhodopsin responds to blue light and produces neuronal firing, while halorhodopsin responds to yellow and silences the cell. Because halorhodopsin also responds to red or near infrared, and because infrared can pass more deeply into tissue, optogenetics offers the possibility of providing nonsurgical control over circuits deeply located in the brain. Optogenetics—a relatively new technology—has been used to study many facets of neuroscience, such as brain reward circuits and mechanisms of memory. It has the potential to supplant deep brain stimulation as a treatment for diseases such as Parkinson's disease or depression, and its potential efficacy for restoring or improving vision by activating damaged retinal cells is also being examined.

Improving Target Identification and Drug Screening

Although disease symptoms are treated as the classic target in drug development, it is obvious that disease mechanisms are a more appropriate and efficient target. Imaging offers a variety of opportunities to improve target identification.

An obvious example of the potential utility of imaging tools is to apply the many methods available from standard imaging technologies such as structural magnetic resonance (sMR), diffusion tensor imaging (DTI), magnetic resonance spectroscopy (MRS), or positron emission tomography (PET) and to use their various measurements to conduct case-control comparisons and to thereby identify anatomic, biochemical, or physiological indicators and mechanisms of disease onset or progression. For example, this approach has been helpful in studying the mechanisms of both classical Mendelian diseases such as Huntington's (which displays evidence of tissue pathology prior to clinical onset that is correlated with the number of CAG repeats) and non-Mendelian diseases such as schizophrenia (which also has indications of tissue change prior to onset that have been linked with increasing replicability to a group of candidate genes such as DISC1, NRG1, and BDNF). Although both of these diseases have evaded treatments that prevent or reverse their onset and progression, imaging research can be used to develop methods that point to new treatment targets. MRS studies of schizophrenia have, for example, supported the search for drugs that affect the glutamate system.

Testing of new compounds that are potentially therapeutic has traditionally been done using animal models and measuring behavioral or metabolic changes. Imaging offers the opportunity to improve on this methodology by offering an opportunity to conduct high-throughput studies of a variety of animal models and also to study disease mechanisms. Mouse and rat models are available for a variety of diseases, and finding new and/or improved models is an important challenge. Another important challenge is to find ways to use imaging to test compounds in simpler systems that can be studied more efficiently and inexpensively. For example, the zebrafish has become an important vertebrate animal model for a variety of human diseases because it is relatively easy to modify genetically. There are now many zebrafish models of diseases, ranging from porphyria to hypothyroidism to age-related cognitive decline. Zebrafish are also susceptible to carcinogens and to infectious agents such as tuberculosis. Therefore, zebrafish offer an attractive option for high-throughput screening of drugs using imaging technologies because such studies can be conducted rapidly and on a large scale. Chemical libraries or potential therapeutic agents could be tested for efficacy or toxicology using zebrafish embryos or larvae and applying digital imaging methods to measure outcome. Investigators are also exploring ways to use more standard imaging technologies (e.g., MR) to screen drugs in mammal animal models.

Key Questions

- How can we be certain that novel imaging methods are yielding valid measures (e.g., the extent to which a specific level of ligand displacement/receptor occupancy reflects optimal treatment levels, the accuracy of tumor volume estimate)? What are their technical or statistical limitations?
- How can existing imaging technologies be applied or modified in novel ways to develop new or better treatments?
- To what extent could new imaging methods such as dual-wavelength laser speckle imaging (measures blood flow, blood volume, and tissue hemoglobin oxygenation) or digital-frequency-ramping optical coherence tomography (images quantitative 3-D vascular network) add new insights to functional imaging?
- What characteristics of an imaging system are most important either for administering therapy or assessing its efficacy?
- What are some ways that high-throughput imaging for drug screening could be enhanced?

Reading

Gradinaru V, Mogri M, Thompson KR, Henderson JM, Deisseroth K. Optical deconstruction of Parkinsonian neural circuitry. *Science* 2009;324:354-9. Accessed online June 15, 2010.

Lieschke GL, Currie PD. Animal models of human disease: zebrafish swim into view. *Nat Rev Genet* 2007;8:253-367. Preview accessed online June 15, 2010.

Mamo D, Kapur S, Shammi CM, Papatheodorou G, Mann S, Therrien F, Remington G. A PET study of dopamine D2 and serotonin 5-HT2 receptor occupancy in patients with schizophrenia treated with therapeutic doses of ziprasidone. *Amer J Psychiatry* 2004;161:818-25. Accessed online June 15, 2010.

IDR TEAM SUMMARY 7   83

> *Because of the popularity of this topic, three groups
> explored this subject. Please be sure to review the second and
> third write-ups, which immediately follows this one.*

IDR TEAM MEMBERS—GROUP A

- Robert S. Balaban, National Institutes of Health
- Shelley A. Batts, Stanford University
- Ashley Grant, University of Texas
- Cindy M. Grimm, Washington University in St. Louis
- Joseph J. Jankowski, Case Western Reserve University
- Mark W. Lenox, Texas A&M University
- Anant Madabhushi, Rutgers University
- Amina A. Qutub, Rice University
- Lyudmila A. Sakhanenko, Michigan State University
- Kimani C. Toussaint, Jr., University of Illinois at Urbana-Champaign
- Roma Subramanian, Texas A&M University

IDR TEAM SUMMARY—GROUP A

*Roma Subramanian, NAKFI Science Writing Scholar,
Texas A&M University*

It is the year 2030. Wanda goes in for her annual mammogram, which is performed using monochromatic X-ray imaging. The image reveals a breast tumor. Wanda's DNA profile reveals mutations in tumor-suppressor genes BRCA1/2, indicating increased breast-cancer risk. Image-guided biopsy is then performed using multimodal imaging, which provides both functional and anatomical data. The resulting imaging data together with Wanda's genomic data are fed into a national systems-medicine database to identify possible treatments and their probable outcomes. Minimally invasive image-guided surgery is selected to resect the tumor. Further, an inoperable metastasis detected in the lung is treated by targeted drug administration and real-time image-guided evaluation of the efficiency and efficacy of drug delivery.

After two days of lively brainstorming at the 2010 National Academies Keck *Futures Initiative* Conference on Imaging Science, an interdisciplinary team of nine researchers, with backgrounds ranging from electrical engineering to infectious disease, began their presentation of the solutions to the challenge posed to the them with the above narrative, which presents a futuristic vision of medical imaging.

As the team members acknowledged, achieving this vision rests on three foundations: creating improvements in imaging technology, diagnostics, and therapeutics.

IDR team 7A began tackling the challenge by drawing up the following list of novel imaging technologies and their applications in diagnostics and therapeutics.

Novel Imaging Technology

- <u>Two-photon fluorescence microendoscopy</u>: This minimally invasive imaging technology provides micron-scale resolution images of tissues in regions inaccessible by light microscopy. It combines two-photon fluorescence microscopy, which eliminates light scatter from deep tissues, and microendoscopy, which enables the visualization of deep-seated structures through the use of endoscope probes composed of microlenses.

  A clinical application of this technology is in the surgical technique of cochlear implant electrode insertion. Handheld portable fluorescent microendoscopes have been developed to visualize the middle and inner ear during this procedure, which is currently being done almost blind. Cochlear sensory cells that are destroyed during the process of electrode insertion result in the loss of any residual hearing. By enabling visualization of the cochlear space, the microendoscope will enable more precise location of the electrode and increase the electrode-nerve interface of cochlear implants.

- <u>Multimodality imaging</u>: An ideal imaging system would be able to simultaneously provide anatomical, physiological, biochemical, and molecular information with high sensitivity and specificity. However, currently, no such single imaging system exists. Different imaging modalities provide different types of information. For example, although magnetic resonance imaging provides high-resolution anatomical images (for example, of brain regions and the connections between them), it cannot image human molecular events. With positron emission tomography, on the other hand, biochemical processes such as protein synthesis and amino acid transport can be observed. Through multimodal imaging, the strengths of each of these imaging systems can be combined.

Integrating in a quantitative manner information from various imaging modalities will help better predict patient outcome. The team also discussed the importance of integrating information from not only various imaging modalities but also from animal models and -omics technology (such as genomics, proteomics, and metabolomics) to aid diagnostic, prognostic, and theranostic predictions and to serve as a training tool.

- <u>Monochromatic X-ray imaging</u>: Phase-contrast and fluorescence monochromatic X-ray imaging have the potential to provide nanometer-scale resolution images of deep tissue at a considerably lower radiation dose than conventional X-rays and without the use of ionizing radiation, which is known to cause DNA damage. Medical applications of this technology include obtaining "freeze" organ motion images, for example, of breathing or of the heart beat; imaging tumors, the edges between organs, and internal structures of bones; and enhancing drug dosing and delivery.

Currently, monochromatic X-rays are generated in a synchrotron, and it is the size and the cost of this machine that are the major barriers to the dissemination of monochromatic X-ray imaging technology. Table-top monochromatic X-ray sources have not yet been developed. However, if this technical challenge could be overcome and a moderately priced, reasonably sized monochromatic X-ray source could be developed, it would revolutionize clinical imaging.

- <u>Photoacoustic imaging</u>: This novel technology is based on the photoacoustic effect, that is, the generation of sound from light (the conversion of non-ionizing laser pulses into ultrasound waves). It combines the resolution provided by ultrasound waves with the high contrast provided by light waves to generate images of deep structures in the body without any health risk. This technology can be used to image blood vessels or tumors or to guide biopsies.

Other imaging technologies the team touched upon included wavefront engineering (with applications in deep-tissue imaging) and mass spectrometry imaging (for imaging and mapping biomolecules in tissue sections, for example, for identifying biomarkers for early cancer diagnosis).

Novel Imaging Applications for Diagnostics and Therapeutics

1. <u>Monitoring drug treatment efficacy</u>: The team discussed the potential of imaging technology for monitoring drug delivery in real time. Applications of image-guided real-time drug delivery include verifying drug delivery on target, monitoring drug release and treatment effects, identifying novel drug targets, determining appropriate drug dosage, and comparing drug treatments.

2. <u>Image-guided surgery</u>: Image-guided therapeutic interventions result in better treatment outcomes by minimizing patient impact and reducing recovery time. For example, the team discussed how minimally invasive image-guided mitral valve repair has the potential to reduce complications associated with open-heart surgery and cardiopulmonary bypass.

For image-guided surgery to improve significantly, surgeons should be able to control the three-dimensional field of view intuitively and manipulate surgical images in real-time. Further, surgical instruments and devices that can be seen and tracked by the selected imaging technique, that have sensors for haptic feedback (to enable remote surgery), and that are integrated with imaging technology so that they have the capability to provide high-magnification, small field-of-view images are required.

Improving Image Processing

As a member of the team explained, there are two parts to the problem of improving imaging technology: (1) acquiring imaging data and (2) processing that data to enhance their information content. Therefore, in addition to enumerating new imaging technologies (or new imaging applications), the team discussed current problems in archiving and processing image data.

Creating Image Databases

Currently, there exists no single clinical image database that can be used to share, search, and retrieve medical imaging records.

One requirement for creating such a database is to store images in an identical format. DICOM—Digital Imaging and Communication in Medicine—is a standard imaging format like "jpg" or "tiff," created to enable the exchange of digital image information between imaging instruments from various vendors. However, DICOM is not a true standard, and inter-vendor operability continues to be a challenge (that is, there are variations in image format derived from machines manufactured by different vendors because there are variations in the way each vendor conforms to the DICOM standard).

Another issue in creating such databases is tracking the provenance and manipulation of imaging data. For example, information about the subject, machine-specific settings or parameters used to acquire the image, and how the image was processed is often unorganized and stored in files in different machines, making it difficult to reanalyze data, assess the quality or usefulness of the imaging data, or replicate experiments.

Other issues discussed in the context of image database construction were difficulties in enforcing clinical image data disclosure; image annotation; image database encryption, privacy, and security; and the need to create databases that correlate image information with biological mechanisms and that enable cross-referencing of image information provided by different modalities (for example, radiological and histological data).

The team concurred that the "full potential" of imaging lay in being able to share imaging data and discussed some applications of image processing/data mining of large sets of collected images.

- <u>Radiation dose modulation</u>: With regard to prostate cancer, computational image segmentation in conjunction with multiparametric magnetic resonance imaging can help determine the location of the tumor so that high radiation dose can be targeted to the tumor area, thereby minimizing the radiation dose to other areas.
- <u>Personalized medicine</u>: Again, in the context of prostate cancer, multiparametric magnetic resonance imaging along with machine learning or pattern recognition tools can be used to distinguish high- and low-grade prostate cancer patients (that is, determine the stage of the disease). This information can thus be used to triage patients for either surgery or enrolment in an active surveillance (or wait-and-watch) program.
- <u>Statistical atlases for diagnosis and treatment</u>: Statistical 3-D population-based atlases of a particular organ provide statistical information on how the structure and function of that organ vary by age, gender, and disease states in large populations. These atlases are generally constructed by integrating data from multiple subjects from different sources (for example, MRI, PET, histology). They are useful for determining normal variations present in a population. Further, because such atlases provide information

on the spatial distribution of a disease, they can help determine which part of an organ the disease is most likely to occur, which in turn will facilitate disease diagnosis and enable accurate biopsies and targeted treatment.

The team concluded its discussion by recommending the endorsement of the following policies by the National Academy of Sciences.

1. Existing programs that promote the deposition of research data in public repositories should be encouraged. Repository creation, administration, and access should be supported by the National Institutes of Health and enforced by scientific journals. Such repositories will enable the creation of an adaptive systems-medicine database and will be necessary for the statistical analysis of large-scale data. Other advantages of such repositories are better detection of disease subtypes and developing personalized treatment.

2. The development of low radiation, minimally invasive imaging for clinical use should be encouraged, for example, monochromatic X-rays. As discussed earlier, these provide contrast and resolution at lower radiation doses.

3. In keeping with the interdisciplinary nature of the NAKFI conference, grants that encourage and fund scientists in different but complimentary disciplines should be encouraged. Funding mechanisms should be leveraged to encourage interdisciplinary research collaborations among individuals with backgrounds in computation, biology, imaging, and clinical medicine.

4. Existing imaging and information technology should be harnessed to improve global health. For example, thermal imaging and T-rays (terahertz radiation) can be used to screen for infectious diseases at airports to understand mechanisms of disease transmission, and mobile phones can be used to image, share, and review global disease data.

IDR TEAM MEMBERS—GROUP B

- Stephen A. Boppart, University of Illinois at Urbana-Champaign
- Danny Ziyi Chen, University of Notre Dame
- Teng-Leong Chew, Northwestern University
- Ivan J. Dmochowski, University of Pennsylvania
- David S. Lalush, North Carolina State University
- Philip J. Santangelo, Georgia Institute of Technology/Emory University

- Joseph A. Zasadzinski, University of California, Santa Barbara
- Laura Smith, University of Georgia

IDR TEAM SUMMARY—GROUP B

*Laura Smith, NAKFI Science Writing Scholar,
University of Georgia*

Medical imaging is an invaluable tool in the diagnosis and treatment of diseases and has the potential to provide even more extensive insight into development and effectiveness for patients with all kinds of acute and long-term diseases. However, the process of imaging is not as simple as taking a picture of the patient's body, discerning some abnormality, and devising an appropriate treatment plan. Instead, the process is complex and involves varying components that are each essential to the smooth operation of the imaging system and the subsequent health implications for the patient. Understanding each of these elements within the overall process of medical imaging was the challenge presented to a group of interdisciplinary scientists that convened at the National Academies Keck *Futures Initiative* Conference on Imaging Science.

The conference brought together experts in a wide range of disciplines, from biomedical engineering to chemistry. IDR team 7B was charged with tackling the question of how to use imaging to improve the treatment of diseases. During the two days of the conference, the team developed a multi-stage approach to the challenge by taking into account the complex nature of medical imaging and disease pathology.

Finding Focus

The key to finding a solution became apparent within the first few minutes of the initial group meeting: treatment. More specifically, the temporal aspects of treatment were considered to be a major concern and one of the focal points around which the entire model revolved. The care of a patient typically follows a prescribed path, beginning with a diagnosis. A treatment decision is then made and implemented. The efficacy of treatment is assessed after a certain amount of time has passed. In the case of cancers, for instance, treatment assessment may not occur until months after the treatment application. The goal of the team became to establish a model that would effectively shorten, or possibly eradicate, the amount of time between these three stages.

In addition, the team decided to focus on developing a plan to aid in the acquisition of imaging data that could improve the therapy decision-making process, as well as the prediction of treatment outcomes. This is especially important when treating a disease that may require long-term management or that has the ability to metastasize, such as cancer and neurodegenerative disorders.

An Integrated Approach

The outcome of two days of group deliberation was not a quick-fix or even a conclusion that just one aspect of imaging could be improved to aid in disease treatment. What arose from the meetings was similar to a business plan or model. The team honed in on what it considered the most crucial aspects of imaging to create an "integrated platform" that would improve the overall state of medical imaging and its ability to aid in the treatment process. This platform would require extensive research in certain areas and includes innovative concepts related to computation, data integration, and human observers of imaging data, technical and chemical components of instrumentation, and the patient. The platform would be used in a situation in which an abnormality or problem of some kind has clearly been identified rather than during the screening process.

"Technology is still of the snapshot variety."

The team approached the challenge by first discussing the current state of imaging technology and what will need to change to make significant improvement. Positron emission tomography (PET) has become the most efficient method of imaging on the molecular and cellular level, capturing the functional processes within the body, such as metabolic activity. More traditional modalities like magnetic resonance (MR) and computed tomography (CT) image anatomical aspects of the body. To improve the information derived from imaging systems, the team concluded that functional imaging or a hybrid of modalities (PET/CT, for example) will be needed to adequately portray activity possibly related to the patient's disorder. The importance of functional imaging lies in the fact that chemical or biological changes related to a disease process often show up before morphological or anatomical evidence becomes noticeable. Functional imaging also provides better contrast in images, which is crucial in discerning abnormalities, like differences between healthy tissue and abnormal tissue.

Functional imaging requires the use of an agent to create a signal in the targeted area of the body, and the team began to think about novel ways agents could be used to not only produce images, but also to aid in the treatment process as well. What it came up with was the option of a multifunctional theranostic agent, or possibly a set of agents. A theranostic combines an imaging agent with a therapeutic agent, thus combining two steps into one. More specifically, this theranostic would contain several imaging probes, but also a therapeutic agent that could be activated once in the targeted area. Such an agent would be able to provide more definitive contrast between the normal and abnormal tissue or structures, perhaps at an earlier stage in pathology than current probes can. In addition, the number of probes included would ensure activation and a signal at any sign of an abnormality. This "cocktail" of agents could also be personalized to the patient's unique case, making a streamlined process allowing diagnosis and treatment to occur close to one another, reducing the treatment time gap mentioned earlier.

"Do we generate the standard human?"

The team also considered the importance of quantization and computation in imaging, where it is sometimes difficult to obtain quantified information. However, after some debate over the importance of computer analysis in understanding medical images, the team arrived at the idea of large-scale data integration as beneficial to the disease treatment process. Although computers cannot perform all the functions a human observer can, humans have limitations that a computer does not. Computer-aided analysis could potentially change the manner in which imaging and data analysis are performed and used to decide on treatment plans for patients.

To speak to the issue of data integration the team suggested a "library" of medical images to aid in the treatment decision-making process. A massive, image-based catalog (similar to the DNA sequence database) would be available for radiologists to compare a patient's image with other existing images demonstrating disease characteristics, matching it as closely as possible to an image in that database. The scaled, catalogued images would represent the standard of human normality, along with any deviations from that baseline according to diseases and their processes. A radiologist would be able to look at an image and determine how much, in quantitative terms, the target is out of the norm. Although a person would make the final call, having a streamlined, computerized way of helping to determine the closest match to a patient's condition would cut out a large chunk of the time spent on human analysis of imaging data.

The human element

Although quantization of data is an important aspect of medical imaging, the team made sure to retain the human observer as a major, vital component of the framework. Computer-aided analysis may be helpful in speeding up the decision-making process, but in the end a human will decide the meaning of the image data, affecting how it is addressed through a treatment plan. Because of this, human interpretation of data needs to be optimized through training, improved communication, and a specialized decision-maker interface between the quantified data and the human interpreter. By understanding that computers cannot perform certain tasks that a person can, the team remained grounded and focused on finding answers that could be more easily obtained in the near future after the improvement of already existing technologies and methods.

Limitations and Recommendations for Research

The model created by the team was composed of several interrelated layers, so it made a point of identifying all areas to be explored further in order for the model to be a viable framework to follow. It is crucial to understand that this platform cannot go forward without recognizing what is lacking in medical imaging for disease treatment and proposing areas that need further research.

The most pressing issue when it comes to advancements in functional imaging is biological target identification. If the targeted object is not an appropriate biomarker, or indicator of the desired biological process, then the data generated from that image is useless for the purpose of designing a treatment plan. This stems from limited understandings of the biological processes involved in certain diseases. Furthering our knowledge of what biological processes signal the onset or presence and progression of diseases will open several doors, allowing researchers to correctly target activities indicative of disease. Furthermore, current technology lacks the sensitivity and specificity to target these biomarkers, limiting what processes can be imaged. In addition to biomarker identification, suitable multifunctional imaging agents would need to be developed to provide the ingredients for the theranostic cocktail.

Another area of research that must be addressed is that of extensive data synthesis. Currently there are no known efforts to integrate multiple datasets in medical imaging. Data would need to be integrated across time, scales, targets, and modalities, which would be a huge effort requiring extensive funding and organization. Without the compilation of this information, however, the decision-making process and subsequent treatment plans would remain time consuming and the interpretation of data divided. Although computer-aided analysis is an exciting concept for medical imaging, it is also one of the most difficult to put into action effectively. It would require the collaboration of many people, the development of reliable hardware, software, and computational algorithms, and acceptance of which data should be synthesized. Additionally, such integrated data would be meaningless without the human observer to visualize and communicate it to others effectively. This creates the need for a dependable interface with which to interpret and communicate the data, as well as training of those involved.

Although the technical limitations of such a model are numerous and necessary to attend to before implementation, one piece of the overall process should be kept in mind at all times: the patient. As with any medical process, it is important to take into account the needs of the person receiving treatment, as well as the demands such procedures may place on someone suffering from a disease. Human compliance is not only desired, but also necessary for such an integrated platform to aid in the treatment of disease.

How Will This Model Affect Society?

The outcome of the team deliberations is not simply the wishful thinking of scientists and researchers invested in the medical imaging field. The ideas generated during the conference are forward-looking and reflect an aspiration for improvement in the overall quality of medical services. The suggested model would provide patients with more effective, personalized treatments tailored to their specific situations. It could also enhance the quality of life for patients by reducing the amount of time spent undergoing, assessing, and changing treatments. The team further predicts patient outcomes to be improved by such personalized treatment regimens. For doctors and researchers, the benefits are widespread; they will be able to better understand disease pathogenesis and biological/physiological processes of the body. Technologically, the advancements in imaging modalities and computerized data will no doubt have spin-off effects for other areas. In addition, the medical field will benefit from the development of computational simplification and data integration by improving ways of handling large amounts of data on patients and conditions. Although time, cost and manpower may all be issues to consider in the present, the future of medical imaging holds promise for the treatment of disease and ultimately the quality of health care and patient health.

IDR TEAM MEMBERS—GROUP C

- Rigoberto C. Advincula, University of Houston
- Stuart S. Berr, University of Virginia
- Frank Chuang, University of California, Davis
- Allan V. Kalueff, Tulane University
- Philip R. LeDuc, Carnegie Mellon University
- John D. MacKenzie, University of California, San Francisco
- Mark J. Schnitzer, Stanford University
- Andrew Tsourkas, University of Pennsylvania
- Alexander Walsh, University of Southern California
- Andrew Z. Wang, University of Northern Carolina
- Nadia Drake, University of California, Santa Cruz

IDR TEAM SUMMARY—GROUP C

*Nadia Drake, NAKFI Science Writing Scholar,
University of California, Santa Cruz*

How Much Is an Image Worth? A Dozen Tests? A Hundred Days? A Thousand Clinical Trials?

In considering how imaging could be used to improve disease treatment, IDR team 7C chose cancer as a disease model. The team's goals were three-pronged: streamlining diagnostic processes for patients by developing multimodal, multiplexed imaging; improving treatments by identifying imaging markers correlating with good or bad outcomes; and making these proposed technologies inexpensive, portable, and accessible to all patients.

At present it can sometimes take months to diagnose and begin treating cancer. Between the first suspicion of a tumor to initial treatment there may be an arduous diagnostic journey. Patients may require more than one kind of image before the size or spread of a tumor can be determined. Biopsies are usually performed. Then, a treatment plan is prepared, often limited to a standard protocol that generally cannot yet be tailored to the specific patient or to his or her specific tumor. Follow-ups happen every three to six months because technologies aren't sensitive enough to detect small numbers of regrowing tumor cells.

In an ideal world, one envisioned by the IDR team, the process of diagnosing and monitoring tumors could be significantly improved, and made more efficient, if new, highly sensitive imaging technology could identify and monitor minute changes in tumor status or spread.

What if, instead of months, the process took days? A tumor is suspected—and a small, portable instrument with multiple imaging capabilities detects and characterizes the tumor that same day. Treatment is based on a detailed dataset containing outcomes for specific biomarkers within a specific lesion, courtesy of finely detailed images that help clinicians characterize the tumor. Follow-ups happen regularly and at home, with remote, continuous monitoring that is non-invasive and sensitive enough to provide a clear picture of even small changes in a tumor or detect metastatic tumor cells traveling in the blood.

Instruments enabling sensitive and efficient ways to diagnose and treat disease might be on the horizon. Such diagnostic and therapeutic innovations might help not only with cancer, but also other conditions such as neurological disorders and serious infections.

Team 7C

Team 7C comprised scientists of all stripes. One team member uses video tracking to study how zebrafish respond to a variety of drugs. Another attaches tiny microscopes to mice and studies the neural correlates of behavior. A third tracks tumors using immune cells. Another studies nanoprobes, which might be useful as new diagnostic or therapeutic tools. They all have an interest in imaging. They all spent two days pooling their collective experience and imaginations to propose an answer to the team's challenge: Find novel ways to use imaging methods to improve the treatment of diseases.

Initial thoughts on using imaging to improve disease treatment were as varied as cell surface markers.

One idea was to do the inverse of conventional post-treatment tumor imaging and develop an imaging agent that identifies unresponsive cells. Another idea was to develop computers recognizing histological patterns—and biomarkers—in a high-throughput manner. A third suggestion was to develop synthetic cells based on fluorescent cell mapping—and in that way, develop a clearer picture of tumor cells and how they might respond. Yet another suggestion was something similar to the Star Trek tricorder—a hand-held scanner that quickly identifies anomalies.

Finding Common Threads

Common to all these ideas? The marriage of technology and clinical application, and the need for multi-functional detection and imaging systems.

Team members cautioned against developing a new technology doctors can't use or understand, saying clinical utility should be paramount in any design process. "We need to give some kind of meaning to it, get the clinical side to validate an imaging technique," a team member said.

Multifunctional imaging systems will improve the speed with which diseases are identified and diagnosed—and also provide more specific information about what a patient is facing. For example, the reagents used in certain types of imaging—positron emission tomography (PET) scans or magnetic resonance imaging (MRI)—are different. The machines are different. The type of data produced by the images is different.

What if imaging modalities, like MRI and PET, could be combined into a single technology routinely providing both anatomical (MRI) and molecular (PET) information? Dual PET-MRI has just been developed and is not widely used. Although seemingly ideal, combining modalities runs into serious stumbling blocks, including finding plausible contrast agents. Such agents are used to increase the visible distinction between different tissues or structures, and can be injected, ingested, or inhaled.

Technologies like computed tomography (CT) use contrast reagents in the form of dyes typically containing iodine or barium. PET scans use radioactive biological analogs to provide images of functioning tissues. MRI enables visualization of sub-surface structures with the help of paramagnetic gadolinium-containing contrast reagents.

"Can we find a contrast agent that will work for both CT and MRI?" a team member asked. Reagents are needed that can be used either sequentially or in parallel, and although no answer readily presented itself, the team considered the possibility of nanoparticles fitting the bill.

Multimodal, Multiplexed Imaging

What should be considered for multiplexed, multimodal imaging development?

Modalities refer to different imaging systems—those that use different target molecules as sources of data. For example, magnetic resonance records a proton signal. CT scans record X-ray attenuation. PET scans look at electron-positron annihilation events. Ultrasound relies on different tissue densities reflecting sound waves. "Multiple markers might not be a big challenge, but seeing them simultaneously is."

Integrating methods requires simultaneously detecting all the different contrast agents or markers in play and making sense of the data they're producing. That also requires that different markers play well with another.

"Energy conversion systems" might easily lend themselves to multimodal integration. These systems involve using things like light and sound as contrast agents—meaning tissues respond to each element differently, allowing a detector to convert the contrasting signals into a picture. Photoacoustic imaging is one example: when light waves are projected into the skin, some are converted into heat and then into ultrasonic waves, which return a highly detailed image, different from the somewhat fuzzy ultrasound images we are used to seeing (ultrasound sends sound waves into the skin). Piezoelectric detection, another form of energy conversion, takes advantage of related electrical and mechanical properties. Traditionally, mechanical stress applied to a material can induce electrical activity, which can be detected and used for imaging.

Multiplexing probes means using the same mode to measure different markers. In theory, a unique probe could be made for different biological markers—either at a subcellular, cellular, or system level. Then, one detector could be used to image all the probes and provide a more integrated picture than a single probe allows.

For example, if a subcutaneous fluorescent imager could "see" many different colors of injected fluorescent markers—each attached to a different type of cell or tissue—then it could weave together a more instructive image than one simply looking at, for example, green-tagged ovarian tumor cells.

Two or three or five multiplexed probes could help researchers extract a lot of information from a single imaging procedure. For instance, if yellow-tagged blood vessels were in the image along with markers of the tumors's genes, then physicians could simultaneously study tumor vascularization and genetics. If blue-tagged epithelial cells were visible, then physicians would know whether the mass was epithelial in origin—and likely cancerous. If orange-tagged germ cells were detectable, then clinicians would know whether the mass might be a benign germ cell tumor. Combing many probes could provide information about which biomarkers are associated with cancerous growth and enable correlations between biomarker presence and clinical outcome. It would be like a cocktail of different probe molecules, each with a unique signature.

Team members also considered whether CT scans could identify different nanoparticle densities—or even make use of Mossbauer probes.

In essence, these proposed imaging technologies would facilitate a level of analysis approaching in vivo cytology and provide an instant, accurate picture of what's going on beneath the skin's surface.

And Then?

Then, that information could be included in detailed datasets about how different biomarkers—identified by different probes—correlate with treatment outcomes. Morphology, proteins, cell cycle alterations, apoptosis, immortality, location, and gene expression and sequence could all be monitored. Instead of two-dimensional datasets—like those considering drug dose and response—team members proposed including the above markers and additional variables like age, sex, time, treatment, genetic information, and environmental factors. This way, imaging could enable more tailored therapies, instead of the one-size-fits-all generic treatment blanket currently covering treatment options. Are we going to reach the level of personalized medicine? Maybe one day. But for now, more specificity would be a good start.

Post-treatment monitoring of metastatic potential could use noninvasive imaging systems that track labeled tumor cells, for example, before they begin cancerous regrowth. The group suggested that constant monitoring—on a detailed scale—is necessary for both the patient's and the clinician's peace of mind. New multimodal monitoring of blood vessels for metastatic tumor cells would offer a kind of constant vigilance that would be more reassuring and medically beneficial than the staggered, routine three- to six-month follow-ups made necessary by current limitations in finding tiny numbers of migrating cells.

The Portability Factor

In addition to simply creating technologies allowing multiple imaging modalities and the monitoring of specific biomarkers, team 7C considered the accessibility of such technologies.

Ideally, instruments would be small. Portable, even. And inexpensive.

Like the Star Trek tricorder, for example—a hand-held device used to scan the body and detect aberrations. What if something like that existed? Something not much larger than a cell phone, with an image screen, that could detect tumors on the spot? Instead of suspecting a tumor—and, in some cases, having no idea where it is—the "tricorder" could assist in determining location, cell type, and prognosis.

Team 7C agreed such a device would be ideal—and maybe even possible. But there are substantial technological hoops to jump through. And developing it would take time. At least 25 years. It's not on the immediate horizon, but within the realm of possibility.

Post-treatment monitoring could be accomplished by equally non-invasive, portable devices: think cell phones, goggles, transcutaneous patches, and fingertip scans—like those used to measure blood oxygenation today. What if organs, tumors, or cells were labeled with a detectable marker, and a smart phone app could turn the phone into a detector? Post-treatment monitoring for tumor regrowth would be possible at home. And results could be text-messaged to a clinician.

Similarly, monitoring the blood for unwelcome travelers (metastatic cells) could be done by taking advantage of the eye's or skin's relative transparency. 3-D goggles could simultaneously scan retinal blood vessels and provide an exciting cinematic experience (Star Trek in 3-D?). A transcutaneous patch or fingertip scanner could be worn overnight and provide data in the morning about whether anything unwanted is running around in the blood. And, in theory, these types of remote monitoring instruments might be possible in about a decade.

The Final Act

Every journalist is familiar with the inverted pyramid—and team 7C is, too. In fact, so are most conference attendees. While giving the team's preliminary report, Andrew Tsourkas pointed to an inverted pyramid—it held broad symptoms of cancerous lesions on top, and drilled down to molecular specifics at its point. The point was to represent new ways of thinking about disease diagnosis and treatment and blurring the lines generally separating levels of disease characterization. Alas, Andrew was accused of not knowing what a pyramid looks like, as the one the group developed was situated on its head.

But team 7C was ready to defend itself: "Building a pyramid right is easy. Building it upside down is impossible. Do the impossible," they said. "Think big."

IDR Team Summary 8

Develop image-specialized database tools for data stewardship and system design in large-scale applications.

CHALLEGE SUMMARY

During the past 30 years imaging science has produced a wide array of image acquisition systems that have revolutionized our ability to acquire images. For example, the evolution from CCD (charge-coupled device) imagers to CMOS (complementary metal oxide semiconductor) imagers has made the acquisition of visible band images nearly free; still and video images of the natural environment and social groups are being acquired at an unprecedented rate. These are being used for mobile visual search applications, in which users acquire cell phone images to navigate their local environment. Medical images in both research and clinical applications, including CT, PET and MR, are being acquired at a rate that is hard to imagine. The diagnosis of such images can be greatly improved by aggregation of datasets.

The revolution in imaging applications has been led by instrumentation—the development of new sensors and data storage technologies that acquire and store many gigabytes of data. Unfortunately, there is not a corresponding effort to develop software database tools to manage this flood of data, and imaging systems are not typically designed with both the hardware and software in mind. For example, because of the nature of the instrument-led acquisition, only a modest amount of information about the imaging context (often called metadata) was planned as part of the instrument design. Moreover, there are no widely accessible tools for aggregating the images and the modest amount of metadata to expand our understanding of natural phenomenon. The aggregation of these data can have applications in a wide range of fields including law, education, business, and medicine.

There is an opportunity—and a need—to design imaging systems from the ground up, keeping both hardware and software in mind. The systems should facilitate the validation, preservation, and analysis of massive amounts of data. For example, the next generation of MR scanners should incorporate the software design team in the first stages of system planning, and the instruments should be engineered for the Exabyte scale. This type of engineering will require the cooperation of research scientists spanning the imaging community and software communities; these individuals typically have very different skill sets and are trained in different university or corporate programs.

Key Questions

- What would it take to build a software infrastructure so that imaging systems developers can easily incorporate large-scale data sharing and data analysis, thereby enabling important information to be coordinated within/among a large user group?
- Are there successful models, such as databases for face recognition and finger printing, that might be used as a model for other organizations, such as MR anatomical and functional data?
- Are there common architectural and computational needs across multiple types of imaging modalities for storing, validating quality, and analyzing image databases? Are there general ontologies for imaging data that might be derived from the images themselves, rather than by labels added by the users in the metadata?

Reading

Brown MS, Shah SK, Pais RC, Lee YZ, McNitt-Gray MF, Goldin JG, Cardenas AF, Aberle DR. Database design and implementation for quantitative image analysis research. *IEEE Trans Inf Technol Biomed* 2005 Mar;9(1):99-108. Accessed online June 15, 2010.

Marcus DS, Archiw KA, Olsen TR, Ramarathnam M. The open-source neuroimaging research enterprise. *J Digital Imaging* epub 2001 Aug 21; Suppl 1:130-8. Accessed online June 15, 2010.

Small SL, Wilde M, Kenny S, Andric M, Hasson U. Database-managed grid-enabled analysis of neuroimaging data: the CNARI framework. *Int J Psychophysiol* epub 2009 Feb 20, 2009 Jul;73(1):62-72. Abstract accessed online June 15, 2010.

IDR TEAM MEMBERS

- Marna E. Ericson, University of Minnesota
- Antonio Facchetti, Polyera Corporation/Northwestern University
- Thomas J. Grabowski, Jr., University of Washington
- Brian P. Hayes, *American Scientist*
- Myrna E. Jacobson Meyers, University of Southern California
- Blake C. Jacquot, Jet Propulsion Laboratory
- Robert H. Lupton, Princeton University
- Rosalind Reid, Harvard University
- Thomasz F. Stepinski, University of Cincinnati
- Tanveer F. Syeda-Mahmood, IBM Almaden Research Center
- Emily Elert, New York University

IDR TEAM SUMMARY

Emily Elert, NAKFI Science Writing Scholar, New York University

Databases, Past and Present

Long before parallel processing, supercomputers, or Turing machines, there were Harvard Computers. These image processors were essential to the telescopic-spectrometry boom of late 19th century astronomy, when new technology was generating information-rich photographs faster than astronomers could analyze them—and before they knew just what they were looking for.

Of course, the Harvard Computers weren't quite like the ones we have today—they were, in fact, a group of women, hired by the astronomer Edward Charles Pickering to process astronomical data. Just as today's computers analyze images and extract meaningful information, Pickering's team went through one glass-plate photograph at a time identifying, measuring, and recording what they saw in the stars.

And it worked! In 1908, after 15 years of this work, Henrietta Swan Leavitt published a paper called "1777 variables in the Magellanic Cloud," which noted a relationship between variable stars' period and luminosity. That discovery, confirmed by Leavitt a few years later, helped set the stage for Hubble's famous red-shift and the understanding that the universe is expanding.

The development of digital imaging has allowed astronomers to acquire tremendous amounts of visual information and rendered analog image processing infeasible. Today, human power is devoted to training computers to identify, measure, and record meaningful information. Rather than hand-written data tables, astronomers organize those extracted features in relational databases, where they can easily be retrieved and analyzed.

This method of image database creation has allowed for some extraordinary scientific investigations. One recent example is the Sloan Digital Sky Survey, in which a dedicated telescope photographed over a quarter of the night sky and catalogued more than 350 million celestial objects. The resulting dataset has yielded some profound discoveries, including the universe's most distant quasars and large populations of sub-stellar objects.

One of the keys to the success of this modern database system is that the physical universe is largely familiar to astronomers, despite its many mysteries. The dataset from the Sloan Survey can be used to nearly perfectly reconstruct images of the sky, because astronomers were able to tell the computer just what they were looking for—they were able to define, in sharp, numerical terms what might constitute meaningful information.

But the modern database system doesn't meet the needs of other, less established sciences. Despite some huge advances in neuroscience and neuroimaging, for example, scientists still lack a basic conceptualization of the structure and function of the brain. Without this understanding, it's often impossible to predict and describe which information in an image of the brain will be useful. Without that ability, it is difficult or impossible to extract all of the relevant features from brain images. Modern imaging database systems can't accommodate the needs of scientists working in fields with these kinds of limitations.

Database Future

The current challenge, then, is figuring out how to acquire imaging data and build databases within rapidly evolving scientific domains. That's a big challenge, but there are a couple of straightforward first steps. In neuroscience, the first step is to standardize the data, both within and across imaging modalities.

Brain imaging technologies are evolving along with scientists' understanding of the brain. Currently, there are no broadly accepted standards in neuroscience for imaging systems and images. Two sets of brain fMRI data from two different studies often yield images taken at different angles with different instrument settings, and then recorded in different file formats with different metadata, and organized into different relational databases.

The result is two bodies of data that have no use beyond the scope of the particular study they were gathered for. It's also quite difficult for other scientists to reproduce their colleagues' findings—a basic practice for the progress of any science.

Standardizing the data would solve both of these problems. Similar standards have been adopted in other fields of imaging and could serve as a model. One of these is Digital Imaging and Communications in Medicine, or DICOM, a standard developed in the 1980s to standardize file formats and metadata. DICOM allows medical images acquired at different places to be transferred and pooled in collective databases.

Another tractable—if more difficult—standardization challenge is that neuroscience imaging operates in a number of modalities. While fMRI uses changes in blood flow as a proxy for brain activity, EEG measures the electrical activity in the brain. MEG, another modality, isolates electromagnetic activity. There's also PET. . . . Each of these modalities has its own strengths and weaknesses, and arguably the field of neuroscience would benefit if there were ways to integrate heterogeneous data across modalities. Ideal databases would be able to pool, weight, and analyze these disparate data to take advantage of the insights each modality can provide.

Creating databases to collect and analyze this data will require a deeper reimagining than the steps outlined so far. Nascent imaging sciences would benefit from databases that can learn, adapt, and change along with the science, and along with evolving imaging technologies. In short, younger sciences require smarter, more agile databases.

These next-generation databases would be tools for exploration as well as analysis. In order to make that possible, images need to become a functional part of the database, along with the numerical features that describe them. The databases need to be able to process images. They need built-in tools for browsing and searching images, and those tools need to be tailored to different scientific domains. In such a database, a user could browse images, select a visual aspect of a single image, and run a search for similar aspects in other images. This is similar to feature extraction, except that the users doesn't have to know exactly what they are looking for—they don't have to be able to define queries in exact, mathematical terms—in order to look.

Those exploratory tools should incorporate machine learning where possible. For example, if a user selects a visual aspect of an image and says, "show me more like this," the computer can return a few results for relevancy feedback from the user. The user can say, "No, not like *this* one—find ones like *this!*" This sort of relevancy feedback can help the user define his or her question, while helping the computer develop more accurate search capabilities.

Currently, the process of feature extraction is limited to database creation. In next-generation databases, feature extraction and imaging data analysis would be an ongoing process. The structure of the relational database would therefore change over time, to reflect evolving scientific understanding.

Recommendations

1. The neuroscience community must define standards for acquiring imaging data and demand that instrument vendors accommodate those standards. Those standards would anticipate the needs of basic science, including:
    a. sharing and searching heterogeneous imaging data;
    b. metadata standards native to instrumentation and specific to neuroscience aims; and
    c. community benchmarks, or ground truth datasets for assessing and stimulating algorithm performance.
2. Scientists must get over their data sharing issues and adopt an open-source model rather than a competitive one.
3. Although the technologies already exist for next-generation databases, the databases themselves do not. Perhaps the biggest reason for this is the lack of interdisciplinary action between people with deep knowledge in a scientific field and people with deep informatics knowledge. Because the problems with current databases have obvious solutions, they fail to interest people in informatics. And because universities reward active research over interdisciplinary expertise, few scientists within those domains have the expertise. In order to create the kind of next-generation databases described here, there must be more interaction between these two groups.
    a. Research is needed into how to pool, evaluate, weight, and use heterogeneous image data.
    b. A plug-in model for database query is desirable, i.e., native support for image processing in the database that has an open modular architecture.
    c. Agile exploratory tools that incorporate image analysis and machine learning must be imagined and implemented for imaging databases.

Appendixes

List of Imaging Science Webcast Tutorials

Stochastic Models of Objects, Images and Imaging Systems
Webcast Released: September 9, 2010
Harrison H. Barrett
Regents Professor
University of Arizona

Task-Based Assessment of Image Quality
Webcast Released: September 9, 2010
Matthew A. Kupinski
Associate Professor
University of Arizona

Imaging Exoplanets
Webcast Released: September 16, 2010
Peter R. Lawson
Chief Technologist, NASA's Exoplanet Exploration Program
Jet Propulsion Laboratory, California Institute of Technology

Statistical Image Models: Engineering, Perception and Neurobiology
Webcast Released: September 16, 2010
Eero P. Simoncelli
Professor of Neural Science, Mathematics and Psychology
New York University

*Using Challenge Problems to Advance the Development of Face Recognition
    Algorithms*
Webcast Released: September 23, 2010
P. Jonathon Phillips
Electronic Engineer
National Institute of Standards and Technology

Multimodal Functional Neuroimaging
Webcast Released: October 7, 2010
Bruce R. Rosen
Professor of Radiology
Harvard Medical School
Professor of Health Services and Technology
Harvard Medical School—Massachusetts Institute of Technology,
    Division of Health Sciences and Technology

Principles of Adaptive Optics
Webcast Released: October 28, 2010
Richard G. Paxman
Chief Scientist and Founder
Diversity Imaging Department
General Dynamics—Advanced Information Systems

All tutorials are available at www.keckfutures.org.

Agenda

Wednesday, November 17, 2010

7:00 and 7:15 a.m.	Bus Pickup: Attendees are asked to allow ample time for breakfast at the Beckman Center; no food or drinks are allowed in the auditorium, which is where the welcome and opening remarks take place at 8:30.
7:30 a.m.	Registration (not necessary for individuals who attended Welcome Reception)
7:30—8:30 a.m.	Breakfast
8:30—8:45 a.m.	**Welcome and Opening Remarks** Harvey V. Fineberg, President, Institute of Medicine Farouk El-Baz, Chair, NAKFI Steering Committee on Imaging Science
8:45—9:45 a.m.	**Keynote Address** Kyle Myers, Director, Division of Imaging and Applied Mathematics, Office of Science and Engineering Laboratories, Center for Devices and Radiological Health, Food and Drug Administration

9:45—10:00 a.m.	**Interdisciplinary Research Team and Grant Program Overview** (Imaging Science Steering Committee Chair)
10:00—10:30 a.m.	Break

Poster Session A Setup

IDR Team Challenge Starters Meet to Review Assignments

10:30 a.m.—12:00 p.m.	Poster Session A

Graduate Science Writing Students to Meet with Barbara Culliton

12:00—1:00 p.m.	Lunch
1:00—5:00 p.m.	Interdisciplinary Research Team Challenge Session 1
3:00—3:30 p.m.	Break

Poster Session B Setup

5:00—7:00 p.m.	Reception/Poster Session B
7:00 p.m.	Bus Pickup: Attendees brought back to hotels

Thursday, November 18, 2010

7:00 and 7:15 a.m.	Bus Pickup
7:15—8:00 a.m.	Breakfast
8:00—10:00 a.m.	Interdisciplinary Research Team Challenge Session 2
10:00—10:30 a.m.	Break

10:30 a.m.—noon	Interdisciplinary Research Team Challenge Reports (5 to 6 minutes per group)
Noon—1:30 p.m.	Lunch Graduate Science Writing Students Meet with Barbara Culliton at Registration Desk for Lunch
1:30—5:00 p.m.	Interdisciplinary Research Team Challenge Session 3
3:00—3:30 p.m.	Break
	Poster Session C Setup: Attendees to set up posters for 5:00 p.m. poster presentation and reception
5:00 p.m.	Final Presentation Drop-Off: Interdisciplinary Research Teams to drop off presentations at information/registration desk, or upload to FTP site prior to 7:00 a.m. Friday morning. (http://nakfi.org/upload, password: upload)
5:00—7:00 p.m.	Poster Session C and Reception All attendees are asked to stop by the registration desk to arrange for last-day transportation if prearranged service does not work with schedule.
7:00 p.m.	Bus Pickup: Attendees brought back to hotel

Friday, November 19, 2010

7:00 and 7:15 a.m.	Bus Pickup: Attendees who are departing for the airport directly from the Beckman Center are asked to bring their luggage to the Beckman Center. Storage space is available.
7:15—8:00 a.m.	Breakfast
7:15 a.m.	Taxi Reservations: Attendees are asked to stop by the information/registration desk to confirm their transportation to the airport or hotel.
8:00—9:30 a.m.	Interdisciplinary Research Team Challenge Reports (8 to 10 minutes per group)
9:30—10:00 a.m.	Break
10:00—11:00 a.m.	Interdisciplinary Research Team Reports (8 to 10 minutes per group)
11:00 a.m.—noon	Q&A Across All Interdisciplinary Research Teams
Noon—1:30 p.m.	Lunch (optional)
Noon—4:00 p.m.	Graduate Science Writing Students Meet with Barbara to Finalize First Draft of Paper

Participant List

Rigoberto C. Advincula
Professor
Chemistry
University of Houston

Sima Bagheri
Professor
Civil & Environmental
 Engineering
NJ Institute of Technology

Richard A. Baird
Director, Division of
 Interdisciplinary Training
National Institute of Biomedical
 Imaging and Bioengineering
 (NIBIB)
National Institutes of Health (NIH)

Chandrajit L. Bajaj
Professor of Computer Sciences,
Comp. App. Mathematics Chair in
 Visualization
Computer Science & Mathematics
University of Texas at Austin

Robert S. Balaban
Scientific Director
National Heart, Lung, and Blood
 Institute
National Institutes of Health

Harrison H. Barrett
Regents Professor of Optical
 Sciences
Professor of Radiology
Professor of Applied Mathematics
College of Optical Sciences and
 Department of Radiology
University of Arizona

Robert J. Barretto
Biochemistry and Biophysics
Columbia University

Randy A. Bartels
Associate Professor
Electrical and Computer
 Engineering
Colorado State University

Mark Bathe
Samuel A. Goldblith Assistant
 Professor of Applied Biology
Biological Engineering
Massachusetts Institute of
 Technology

Shelley A. Batts
Postdoctoral Scholar
NIH NRSA Fellow
Otolaryngology and Bio-X
Stanford University

Stuart S. Berr
Professor of Research
Radiology
University of Virginia

Thomas Bifano
Director (and Professor)
Photonics Center (and Mechanical
 Engineering)
Boston University

Ali Bilgin
Assistant Professor
Biomedical Engineering
University of Arizona

Floyd E. Bloom
Professor Emeritus
Molecular and Integrative
 Neuroscience Department
The Scripps Research Institute

Stephen A. Boppart
Professor of Electrical Engineering,
 Bioengineering, and Medicine
Beckman Institute for Advanced
 Science and Technology
University of Illinois at
 Urbana-Champaign

Liliana Borcea
Noah G. Harding Professor
Computational and Applied
 Mathematics
Rice University

DuBois Bowman
Associate Professor
Biostatistics and Bioinformatics
Emory University

Joseph E. Burns
Health Sciences Assistant Clinical
 Professor
Radiological Sciences
University of California, Irvine
 School of Medicine

Jordan Calmes
NAKFI Science Writing Scholar
Massachusetts Institute of
 Technology

Owen T. Carmichael
Assistant Professor
Neurology
University of California, Davis

Supriya Chakrabarti
Professor
Astronomy
Boston University

Danny Ziyi Chen
Professor
Computer Science and Engineering
University of Notre Dame

Teng-Leong Chew
Director of University Imaging
    Resources
Director of Nikon Imaging Center
Cell and Molecular Biology
Northwestern University

Dennis W. Choi
Executive Vice President
Simons Foundation

Frank Chuang
Associate Director for Research
NSF Center for Biophotonics
University of California, Davis
    Medical Center

Miriam Cohen
Postdoctoral Fellow
Department of Cellular &
    Molecular Medicine
University of California, San Diego

Graham P. Collins
Freelance Science Writer/Editor

Richard S. Conroy
Program Director
National Institute of Biomedical
    Imaging and Bioengineering
National Institutes of Health

Barbara J. Culliton
President
The Culliton Group/Editorial
    Strategies

Ivan J. Dmochowski
Associate Professor
Chemistry
University of Pennsylvania

Nadia Drake
NAKFI Science Writing Scholar
Science Communication
University of California, Santa
    Cruz

Charles Elachi
Director
Jet Propulsion Laboratory

Farouk El-Baz
Research Professor and Director
Center for Remote Sensing
Boston University

Emily Elert
NAKFI Science Writing Scholar
New York Univeristy

Alireza Entezari
Assistant Professor
Computer and Information
    Science and Engineering
    Department
University of Florida

Marna E. Ericson
Assistant Professor
Dermatology
University of Minnesota

Antonio Facchetti
Chief Technology Officer and
    Adjunct Professor
Chemistry
Polyera Corporation and
    Northwestern University

Joyce E. Farrell
Executive Director
Stanford Center for Image Systems
    Engineering
Stanford University

James A. Ferwerda
Associate Professor
Munsell Color Science Laboratory,
    Carlson Center for Imaging
    Science
Rochester Institute of Technology

David A. Fike
Assistant Professor
Earth & Planetary Sciences
Washington University

Harvey V. Fineberg
President
Institute of Medicine

Douglas P. Finkbeiner
Professor
Astronomy & Physics
Harvard University

Susan Fitzpatrick
Vice President
James S. McDonnell Foundation

Jason W. Fleischer
Professor
Electrical Engineering
Princeton University

Felice C. Frankel
Research Associate
Systems Biology
Harvard Medical School

Richard A. Frazin
Assistant Research Scientist
Atmospheric, Oceanic and Space
    Sciences
University of Michigan

Kenneth R. Fulton
Executive Director
National Academy of Sciences

Tia A. Ghose
Freelance Science Writer

Eric Gilleland
Project Scientist
Research Applications Laboratory
National Center for Atmospheric
    Research

Michael Glenn Easter
NAKFI Science Writing Scholar
New York Univeristy

Alyssa A. Goodman
Professor
Astronomy
Harvard University

Thomas J. Grabowski
Professor of Radiology and Neurology
Director, Integrated Brain Imaging Center Radiology
University of Washington

Scott T. Grafton
Director, UCSB Brain Imaging Center
Psychology
University of California, Santa Barbara

Ashley Grant
Graduate Student
Experimental Pathology
University of Texas Medical Branch

Cindy M. Grimm
Associate Professor
Computer Science and Engineering
Washington University in St. Louis

Mark A. Griswold
Associate Professor
Director of Magnetic Resonance Physics Research
Radiology
Case Western Reserve University
University Hospitals of Cleveland

George R. Hale
NAKFI Science Writing Scholar
Texas A&M University

Brian P. Hayes
Senior Writer
American Scientist

Daniel P. Holschneider
Associate Professor
Psychiatry & the Behavioral Sciences
University of Southern California

David M. Hondula
Ph.D. Candidate
Department of Environmental Sciences
The University of Virginia

Xiaoping Hu
Professor and Georgia Research Alliance Eminent Scholar in Imaging
Director, Biomedical Imaging Technology Center
Scientific Director, Center for Systems Imaging
Coulter Department of Biomedical Engineering
Biomedical Imaging Technology Center Center for Systems Imaging
Georgia Tech
Emory University

Myrna E. Jacobson Meyers
Assistant Professor (research)
Wrigley Institute of Environmental Science MBBO
University of Southern California

Blake C. Jacquot
Member of Technical Staff
Advanced Detector and Nano Science Technologies
Jet Propulsion Laboratory

Hamid Jafarkhani
Chancellor's Professor
Electrical Engineering and
 Computer Science
University of California, Irvine

Joseph J. Jankowski
Associate Vice President
Technology Management
Research and Technology
 Management
Case Western Reserve University

Andreas Jeromin
Associate Director
Banyan Biomarkers, Inc

Jeffrey Johnston
Senior Program Officer
W. M. Keck Foundation

Marcel Adam Just
D. O. Hebb Professor
Psychology
Carnegie Mellon University

Allan V. Kalueff
Assistant Professor of
 Pharmacology
Pharmacology and Neuroscience
 Program
Tulane University School of
 Medicine

Farzad Kamalabadi
Associate Professor
Electrical and Computer
 Engineering
University of Illinois

Ana Kasirer-Friede
Associate Project Scientist
Medicine/ Hematology-Oncology
University of California, San Diego

Daniel F. Keefe
Assistant Professor
Department of Computer Science
 and Engineering
University of Minnesota

Cristen Kelly
Program Associate
National Academies Keck *Futures
 Initiative*

Olga Khazan
NAKFI Science Writing Scholar
University of Southern California

Jennifer D. T. Kruschwitz
Principal Optical Coatings
 Engineer—President
JK Consulting

Matthew A. Kupinski
Associate Professor
Optical Sciences
University of Arizona

David S. Lalush
Associate Professor
Associate Department Head
Department of Biomedical
 Engineering
North Carolina State University

Linda J. Larson-Prior
Research Associate Professor
Mallinckrodt Institute of
    Radiology
Washington University in St. Louis
    Medical School

Tod R. Lauer
Astronomer
National Optical Astronomy
    Observatory

Lincoln J. Lauhon
Associate Professor
Materials Science and Engineering
Northwestern University

Peter R. Lawson
Chief Technologist
NASA's Exoplanet Exploration
    Program
Jet Propulsion Laboratory

Philip R. LeDuc
Associate Professor
Mechanical and Biomedical
    Engineering, Computational
    Biology, Biological Science
Carnegie Mellon University

Mark W. Lenox
Director of Imaging
Texas A&M Institute for
    Preclinical Studies
Texas A&M University

Rachel Lesinski
Program Associate
National Academies Keck *Futures*
    Initiative

Craig S. Levin
Professor
Radiology, Physics, and Electrical
    Engineering
Stanford University

Zhi-Pei Liang
Professor
Electrical and Computer
    Engineering
University of Illinois at
    Urbana-Champaign

K. J. Ray Liu
Professor and Associate Chair
Electrical and Computer
    Engineering
University of Maryland

Robert H. Lupton
Senior Research Astronomer
Astrophysical Sciences
Princeton University

John D. MacKenzie
Assistant Professor in Residence
Radiology and Biomedical Imaging
University of California, San
    Francisco

Anant Madabhushi
Associate Professor
Biomedical Engineering
Rutgers University

Mohammad H. Mahoor
Assistant Professor
Electrical and Computer
    Engineering
University of Denver

Jonathan J. Makela
Associate Professor
Department of Electrical and
    Computer Engineering
University of Illinois at
    Urbana-Champaign

Giovanni Marchisio
Principal Scientist
R&D
DigitalGlobe

J. Lawrence Marsh
Professor and Director
Developmental Biology Center
University of California, Irvine

Timothy P. McClanahan
Scientist
Astrochemistry Laboratory
NASA Goddard Space Flight
    Center

Carolyn C. Meltzer
William P. Timmie Professor
Chair of Radiology
Associate Dean for Research
    Radiology
Emory University School of
    Medicine

Shalini Misra
Postdoctoral Fellow
Planning, Policy, and Design
School of Social Ecology
University of California, Irvine

Mahta Moghaddam
Professor
Electrical Engineering and
    Computer Science
The University of Michigan

Victoria Morgan
Assistant Professor of Radiology
    and Radiological Sciences
Assistant Professor of Biomedical
    Engineering
Vanderbilt University

Kyle J. Myers
Director
Division of Imaging and Applied
    Mathematics
Office of Science and Engineering
    Laboratories
Center for Devices and
    Radiological Health
U.S. Food and Drug
    Administration

Hamid Najib
IT Program and Support Specialist
National Academies Keck *Futures
    Initiative*

Teri W. Odom
Associate Professor and Dow
    Chemical Company Research
    Professor
Chemistry
Northwestern University

Joseph A. O'Sullivan
Samuel C. Sachs Professor and
    Dean UMSL/WU Joint
    Engineering Program
Electrical and Systems Engineering
Washington University

Gregory M. Palmer
Assistant Professor
Radiation Oncology
Duke University Medical Center

Thrasyvoulos N. Pappas
Associate Professor & Associate
    Chair
Electrical Engineering and
    Computer Science
Northwestern University

Richard G. Paxman
Chief Scientist and Founder
Diversity Imaging Department
General Dynamics - Advanced
    Information Systems

Maria Pellegrini
Executive Director, Programs
W. M. Keck Foundation

P. Jonathon Phillips
Electronic Engineer
Information Technology
    Laboratory
National Institute of Standards and
    Technology

Rafael Piestun
Director, COSI-IGERT
Electrical Computer and Energy
    Engineering and Physics
University of Colorado at Boulder

Lori Pindar
NAKFI Science Writing Scholar
University of Georgia

Robert B. Pless
Associate Professor
Computer Science and Engineering
Washington University

Steven G. Potkin
Brain Imaging Science Center
    Director Director of Clinical
    Research
Professor
Department of Psychiatry and
    Human Behavior
University of California, Irvine

Lisa A. Poyneer
Engineer
Optical Sciences (Physics) and
    Image and Signal Processing
    (Engineering)
Lawrence Livermore National
    Laboratory

Amina A. Qutub
Assistant Professor
Bioengineering
Rice University

Andrew B. Raij
Postdoctoral Fellow
Department of Computer Science
University of Memphis

A. Ravishankar Rao
Research Staff Member
Computational Biology
IBM Research

Kathleen M. Raven
NAKFI Science Writing Scholar
University of Georgia

James M. Rehg
Professor
School of Interactive Computing
Georgia Institute of Technology

Rosalind Reid
Executive Director
Institute for Applied
    Computational Science
Harvard University School of
    Engineering and Applied
    Sciences

Emmanuel G. Reynaud
School of Biology and
    Environmental Science
University College Dublin

James E. Rhoads
Associate Professor of Astronomy
School of Earth and Space
    Exploration
Arizona State University

Bernice E. Rogowitz
Scientist
Texas Advanced Computing
    Center
University of Texas, Austin

Bruce R. Rosen
Professor in Radiology at Harvard
    Medical School
Director, Athinoula A. Martinos
    Center for Biomedical
    Imaging
Department of Radiology
Massachusetts General Hospital

Keith Rozendal
NAKFI Science Writing Scholar
University of California, Santa
    Cruz

Emily C. Ruppel
NAKFI Science Writing Scholar
School of Humanities, Arts, and
    Social Sciences
Massachusetts Institute of
    Technology

Stephen Ryan
President, Doheny Eye Institute
    Grace Emery Beardsley
    Distinguished Professor in
    Ophthalmology
Board Member W. M. Keck
    Foundation

Naoki Saito
Professor of Mathematics
Chair of Graduate Group in
 Applied Mathematics
Mathematics
University of California, Davis

Lyudmila A. Sakhanenko
Associate Professor
Statistics and Probability
Michigan State University

Philip J. Santangelo
Assistant Professor
Biomedical Engineering
Georgia Institute of Technology
Emory University

Suzanne Scarlata
Professor
Physiology & Biophysics
Stony Brook University

Mark J. Schnitzer
Investigator, Howard Hughes
 Medical Institute
Assistant Professor
Department of Biology
Department of Applied Physics
Stanford University

Joshua W. Shaevitz
Assistant Professor
Physics and the Lewis-Sigler
 Institute for Integrative
 Genomics
Princeton University

Eero P. Simoncelli
Investigator, Howard Hughes
 Medical Institute
Professor, Neural Science,
 Mathematics & Psychology
New York University

Laura Smith
NAKFI Science Writing Scholar
University of Georgia

Allen W. Song
Professor of Radiology, Psychiatry
 and Biomedical Engineering
Director, Brain Imaging and
 Analysis Center
Duke University Medical Center

Tomasz F. Stepinski
Professor
Department of Geography
University of Cincinnati

Daniel Stokols
Chancellor's Professor
Department of Planning, Policy,
 and Design and Department
 of Psychology and Social
 Behavior
Program in Public Health and
 Department of Epidemiology
School of Social Ecology
University of California, Irvine

Roma Subramanian
Graduate Science Writing Student
Texas A&M University

Kimberly Suda-Blake
Senior Program Director
National Academies Keck *Futures Initiative*

Tanveer F. Syeda-Mahmood
Research Manager
Multimodal Mining for Healthcare
IBM Almaden Research Center

Phillip Szuromi
Supervisory Senior Editor
Science

Mercedes Talley
Program Director
W. M. Keck Foundation

Demetri Terzopoulos
Chancellor's Professor of Computer Science
Computer Science Department
University of California, Los Angeles

Jerilyn A. Timlin
Principal Member of Technical Staff
Bioenergy and Defense Technologies
Sandia National Laboratories

Derek K. Toomre
Associate Professor
Cell Biology
Yale University School of Medicine

Kimani C. Toussaint
Assistant Professor
Mechanical Science and Engineering
University of Illinois at Urbana-Champaign

Andrew Tsourkas
Associate Professor
Bioengineering
University of Pennsylvania

Remy Tumbar
Research Scientist
Molecular Biology and Genetics
Cornell University

Kamil Ugurbil
Professor
Center for Magnetic Resonance Research
University of Minnesota

Paul Vaska
Scientist
Medical Department
Brookhaven National Laboratory

Rene Vidal
Associate Professor
Center for Imaging Science
Johns Hopkins University

Tom Vogt
Director, Nanocenter
Professor of Chemistry & Biochemistry
USC NanoCenter
University of South Carolina

Alexander Walsh
Assistant Professor of Opthalmology
Keck School of Medicine
University of Southern California

Jessika Walsten
NAKFI Science Writing Scholar
University of Southern California

Brian A. Wandell
Isaac and Madeline Stein Family Professor
Department of Psychology
Stanford University

Andrew Z. Wang
Assistant Professor
Radiation Oncology
University of North Carolina

Gordon X. Wang
Postdoctoral Fellow
Molecular and Cellular Physiology
Stanford University

Lihong V. Wang
Gene K. Beare Distinguished Professor
Department of Biomedical Engineering
Washington University in St. Louis

Paul S. Weiss
Fred Kavli Chair in NanoSystems Science California NanoSystems Institute (CNSI) Director
Distinguished Professor
CNSI and Chemistry and Biochemistry
University of California, Los Angeles

Emily White
NAKFI Science Writing Scholar
Texas A&M University

Patrick J. Wolfe
Associate Professor
School of Engineering and Applied Sciences Department of Statistics
Harvard University
Harvard-MIT Division of Health Sciences and Technology

Curtis Woodcock
Professor
Department of Geography and Environment
Boston University

Joseph A. Zasadzinski
Professor
Chemical Engineering and Materials
University of California, Santa Barbara

Hongkai Zhao
Professor
Mathematics
University of California, Irvine